Study Guide

T0202200

Pure Mathematics Unit 1
for CAPE®

Kenneth Baisden
Charles Cadogan
Sue Chandler
Mahadeo Deokinandan

Great Clarendon Street, Oxford, OX2 6DP, United Kingdom

Oxford University Press is a department of the University of Oxford.
It furthers the University's objective of excellence in research, scholarship,
and education by publishing worldwide. Oxford is a registered trade mark of
Oxford University Press in the UK and in certain other countries

First published by Nelson Thornes Ltd in 2013
This edition published by Oxford University Press in 2014

British Library Cataloguing in Publication Data
Data available

978-1-4085-2039-0

10 9 8

Printed in Great Britain by CPI Group (UK) Ltd., Croydon CR0 4YY

Acknowledgements

Cover photograph: Mark Lyndersay, Lyndersay Digital, Trinidad
www.lyndersaydigital.com
Page make-up and illustrations: Tech-Set Ltd, Gateshead

Thanks are due to Kenneth Baisden, Charles Cadogan, and Mahadeo Deokinandan
for their contributions in the development of this book.

Although we have made every effort to trace and contact all
copyright holders before publication this has not been possible in all
cases. If notified, the publisher will rectify any errors or omissions at
the earliest opportunity.

Links to third party websites are provided by Oxford in good faith
and for information only. Oxford disclaims any responsibility for
the materials contained in any third party website referenced in
this work.

Contents

Section 3 Calculus 1

Introduction

This Study Guide has been developed exclusively with the Caribbean Examinations Council (CXC®) to be used as an additional resource by candidates, both in and out of school, following the Caribbean Advanced Proficiency Examination (CAPE®) programme.

It has been prepared by a team with expertise in the CAPE® syllabus, teaching and examination. The contents are designed to support learning by providing tools to help you achieve your best in CAPE® Pure Mathematics and the features included make it easier for you to master the key concepts and requirements of the syllabus. *Do remember to refer to your syllabus for full guidance on the course requirements and examination format!*

Inside this Study Guide is an interactive CD which includes electronic activities to assist you in developing good examination techniques:

- **On Your Marks** activities provide sample examination-style short answer and essay type questions, with example candidate answers and feedback from an examiner to show where answers could be improved. These activities will build your understanding, skill level and confidence in answering examination questions.

- **Test Yourself** activities are specifically designed to provide experience of multiple-choice examination questions and helpful feedback will refer you to sections inside the study guide so that you can revise problem areas.

- **Answers** are included on the CD for exercises and practice questions, so that you can check your own work as you proceed.

This unique combination of focused syllabus content and interactive examination practice will provide you with invaluable support to help you reach your full potential in CAPE® Pure Mathematics.

1 Basic algebra and functions

1.1 Terminology and principles

Language of mathematics

The language of mathematics is a combination of words and symbols where each symbol is a shorthand form for a word or phrase. When the words and symbols are used correctly a piece of mathematical reasoning can be read in properly constructed sentences in the same way as a piece of prose.

Many of the words used have precise mathematical definitions. For example, the word 'bearing' has many meanings when used in everyday language, but when used mathematically it means the direction of one point from another.

You need to be able to present your solutions using clear and correct mathematical language and symbols.

Symbols used for operators

A mathematical operator is a rule for combining or changing quantities.

You are already familiar with several operators and the symbols used to describe them.

+ means 'plus' or 'and' or 'together with' or 'followed by', depending on context.

For example, $2 + 5$ means 2 plus 5 or 2 and 5,

$\mathbf{a} + \mathbf{b}$ means $\mathbf{a}$ together with $\mathbf{b}$ or $\mathbf{a}$ followed by $\mathbf{b}$.

$-$ means 'minus' or 'take away'.

For example, $2 - 5$ means 2 minus 5 or 2 take away 5.

The operators $\times$ and $\div$ also have familiar meanings.

Symbols used for comparison

The commonest symbol used for comparing two quantities is $=$ and it means 'is equal to'.

For example, $x = 6$ means x is equal to 6.

Some other familiar symbols are $>$ which means 'is greater than' and $\geq$ which means 'is greater than or is equal to'. A forward slash across a comparison symbol is used to mean 'not', for example, $\neq$, which means 'is not equal to'.

Terms, expressions, equations and identities

To use comparison symbols correctly, you need to recognise the difference between terms, expressions, equations and identities.

A mathematical expression is a group of numbers and/or variables

(for example, x) and operators. For example, $2x$, $3 - 2y$ and $\dfrac{5x^2}{3 - 2x}$ are expressions.

The parts of an expression separated by $+$ or $-$ are called terms. For example, 3 and $2y$ are terms in the expression $3 - 2y$

An equation is a statement saying that two quantities are equal in value.

For example, $2x - 3 = 7$ is a statement that reads '$2x - 3$ is equal to 7'.

This statement is true only when $x = 5$

Some equations are true for any value that the variable can take. For example, $x + x = 2x$ is true for any value of x. This equation is an example of an identity and we use the symbol $\equiv$ to mean 'is identical to'. Therefore we can write $x + x \equiv 2x$

Symbols used for linking statements

When one statement, such as $x^2 = 4$, is followed by another statement that is logically connected, for example $x = \pm 2$, they should be linked by words or symbols.

Some examples of words that can be used are:

'$x^2 = 4$ therefore $x = \pm 2$', '$x^2 = 4$ implies that $x = \pm 2$'

'$x^2 = 4$ so it follows that $x = \pm 2$' '$x^2 = 4$ gives $x = \pm 2$'

'$x^2 = 4$ hence $x = \pm 2$'

The symbols $\therefore$ and $\Rightarrow$ can be used to link statements, where $\therefore$ means 'therefore' or 'hence' and the symbol $\Rightarrow$ means 'implies that' or 'gives'.

For example, $2x - 1 = 5$ $\therefore x = 3$ or $2x - 1 = 5$ $\Rightarrow$ $x = 3$

Setting out a solution

> It is important to set out your solutions to problems using correct linking symbols or words.

> It is also important that you explain the steps you take and your reasoning.

The following example shows a way of explaining the solution of the pair of simultaneous equations: $2x + 3y = 1$ and $3x - 4y = 10$

$$
\begin{array}{llll}
& 2x + 3y = 1 & [1] \\
& 3x - 4y = 10 & [2] \\
[1] \times 3 \Rightarrow & 6x + 9y = 3 & [3] \\
[2] \times 2 \Rightarrow & 6x - 8y = 20 & [4] \\
[3] - [4] \Rightarrow & 17y = -17 \\
\therefore & y = -1
\end{array}
$$

Substituting -1 for y in [1] gives $2x + 3(-1) = 1$ $\Rightarrow$ $x = 2$

The solution is $x = 2$ and $y = -1$

Notice that the equations are numbered. This gives a way of explaining briefly what we are doing to combine them in order to eliminate one of the variables.

Exercise 1.1

In each question, explain the incorrect use of symbols and write down a correct solution.

1 Solve the equation $3x - 1 = 5$

$$
\begin{array}{rl}
& 3x - 1 = 5 \\
= & 3x = 6 \\
= & x = 2
\end{array}
$$

2 Find the value of A given that $\sin A° = 0.5$

$$
\begin{array}{rl}
& \sin A° = 0.5 \\
\therefore & A = 30°
\end{array}
$$

Binary operations

A *binary operation* is a rule for combining two members of a set.

For example, we can combine two members of the set of real numbers by addition, subtraction, multiplication or division. We know the rules for these operations, that is $4 + 2 = 6$, $4 - 2 = 2$, $4 \times 2 = 8$ and $4 \div 2 = 2$

We can define other operations. For example, for a and b, where a, b are members of the set of real numbers, $\mathbb{R}$, then a and b are combined to give $2a - b$

We write this briefly as $a * b = 2a - b$

Then, for example, $3 * 7 = 2 \times 3 - 7 = -1$

Properties of operations

An operation, $*$, is *commutative* when $a * b$ gives the same result as $b * a$ for any two members of the set.

For example, addition on the set of real numbers is commutative because

$$a + b = b + a, \quad a, b \in \mathbb{R}$$

Multiplication on the set of real numbers is also commutative because

$$a \times b = b \times a, \quad a, b \in \mathbb{R}$$

However, subtraction is not commutative because, in general,

$$a - b \neq b - a$$

$$\text{e.g. } 3 - 7 \neq 7 - 3$$

Division is also not commutative because, in general,

$$a \div b \neq b \div a$$

$$\text{e.g. } 3 \div 7 \neq 7 \div 3$$

An operation is *associative* when any three members can be combined by operating on either the first two members or the second two members first, that is

$$(a * b) * c = a * (b * c)$$

For example, multiplication on the set of real numbers is associative because $(a \times b) \times c = a \times (b \times c)$, $a, b, c \in \mathbb{R}$
e.g. $(2 \times 3) \times 4 = 2 \times (3 \times 4)$

Addition on the set of real numbers is also associative because $(a + b) + c = a + (b + c)$, $a, b, c \in \mathbb{R}$
e.g. $(2 + 3) + 4 = 2 + (3 + 4)$

However, subtraction is not associative because, in general, $(a - b) - c \neq a - (b - c)$, $a, b, c \in \mathbb{R}$
e.g. $(2 - 3) - 4 = -1 - 4 = -5$ whereas $2 - (3 - 4) = 2 - (-1) = 3$

Division is also not associative because, in general, $(a \div b) \div c \neq a \div (b \div c)$, $a, b, c \in \mathbb{R}$
e.g. $(2 \div 3) \div 4 = \frac{2}{3} \div 4 = \frac{1}{6}$ whereas $2 \div (3 \div 4) = 2 \div \frac{3}{4} = \frac{8}{3}$

An operation, *, is *distributive* over another operation, ◊, when for any three members of the set $a * (b \lozenge c) = (a * b) \lozenge (a * c)$

For example, multiplication is distributive over addition and subtraction on members of $\mathbb{R}$ because

$$a \times (b + c) = ab + ac \text{ and } a \times (b - c) = ab - ac$$

but multiplication is not distributive over division because

$$a \times (b \div c) = \frac{ab}{c}$$

whereas $(a \times b) \div (a \times c) = \dfrac{b}{c}$

so $\quad a \times (b \div c) \neq (a \times b) \div (a \times c)$ unless $a = 1$

Example

An operation * is defined for all real numbers x and y as $x * y = 2x + 2y$

Determine whether the operation * is: **(a)** commutative **(b)** associative.

(a) $x * y = 2x + 2y$ and $y * x = 2y + 2x$

$2x + 2y = 2y + 2x$ because the addition of real numbers is commutative.

∴ the operation * is commutative.

(b) Taking x, y and z as three real numbers,

$(x * y) * z = (2x + 2y) * z = 2(2x + 2y) + 2z = 4x + 4y + 2z$

$x * (y * z) = x * (2y + 2z) = 2x + 4y + 4z$

Therefore $(x * y) * z \neq x * (y * z)$ so the operation is not associative.

Example

(a) For two real numbers, x and y, the operation * is given by
$x * y = x^2 + y^2$

Determine whether the operation is associative.

(b) For two real numbers, x and y, the operation ◊ is given by $x \lozenge y = xy$

Determine whether the operation ◊ is distributive over the operation *.

(a) For any three real numbers, x, y and z,

$(x * y) * z = (x^2 + y^2) * z = (x^2 + y^2)^2 + z^2$

$x * (y * z) = x^2 * (x^2 + z^2) = x^2 + (x^2 + z^2)^2$

∴ $(x * y) * z \neq x * (y * z)$ so the operation is not associative.

(b) For any three real numbers, x, y and z,

$x \lozenge (y * z) = x(y * z) = x(y^2 + z^2) = xy^2 + xz^2$

$(x \lozenge y) * (x \lozenge z) = xy * xz = x^2y^2 + x^2z^2$

∴ $x \lozenge (y * z) \neq (x \lozenge y) * (x \lozenge z)$

so the operation ◊ is not distributive over the operation *.

Closed sets

A set is *closed* under an operation * when for any two members of the set, $a * b$ gives another member of the set.

For example, the set of integers, $\mathbb{Z}$, is closed under addition because for any two integers, a and b, $a + b$ is also an integer.

However, $\mathbb{Z}$ is not closed under division because $a \div b$ does not always give an integer, for example $3 \div 4 = \frac{3}{4}$, which is not an integer.

Identity

If a is any member of a set and there is one member b of the set, such that under an operation, *, $a * b = b * a = a$ then b is called the *identity member* of the set under the operation.

For example, 0 is the identity for members of $\mathbb{R}$ under addition as, for any member a,

$$0 + a = a + 0 = a$$

However, there is no identity for members of $\mathbb{R}$ under subtraction because there is no real number b such that $a - b = b - a$

Also 1 is the identity for members of $\mathbb{R}$ under multiplication, as for any member a,

$$1 \times a = a \times 1 = a$$

However, there is no identity for members of $\mathbb{R}$ under division because there is no real number b such that $a \div b = b \div a$

Inverse

Any member a of a set has an *inverse* under an operation if there is another member of the set, which when combined with a gives the identity.

Clearly, a member can have an inverse only if there is an identity under the operation.

For example, as 0 is the identity for members of $\mathbb{R}$ under addition, then $-a$ is the inverse of any member a,
since $-a + a = a + (-a) = 0$

Also, as 1 is the identity for members of $\mathbb{R}$ under multiplication,

then $\frac{1}{a}$ is the inverse of a since $a \times \frac{1}{a} = 1$,

but there is one important exception to this:

$$\frac{1}{0} \text{ is meaningless.}$$

Example

An operation, $*$, is defined for all real numbers x and y as

$$x * y = \frac{(x + y)}{2}$$

(a) Show that the set is closed under the operation $*$.

(b) Show that x has no inverse under the operation $*$.

(a) When x and y are real numbers, $\frac{(x + y)}{2}$ is also a real number.

Therefore the set is closed under the operation $*$.

(b) For x to have an inverse, there needs to be an identity member, i.e. one member, b, of $\mathbb{R}$ such that $x * b = x$,

i.e. such that $\frac{(x + b)}{2} = x$

Solving this for b gives $b = x$

Now x is any member of $\mathbb{R}$, so b is not a single member and therefore there is no identity member of the set.

As there is no identity, x has no inverse.

Exercise 1.2

1 Determine whether addition is distributive over multiplication on the set of real numbers.

2 The operation $*$ is given by

$$x * y = x^2 y$$

for all real values of x and y.

Determine whether the operation $*$ is:

(a) commutative

(b) associative

(c) distributive over addition.

3 The operation $*$ is given by

$$x * y = \sqrt{xy}$$

for all positive real numbers including 0.

(a) Show that the operation $*$ is closed.

(b) Write down the identity member.

(c) Determine whether each member has an inverse.

4 The operation $\sim$ is given by $x \sim y =$ the difference between x and y for $x, y \in \mathbb{R}$.

(a) Determine whether $\mathbb{R}$ is closed under this operation.

(b) Show that the identity member is 0.

(c) Show that each member is its own inverse.

Surds

The square roots of most positive integers and fractions cannot be expressed exactly as either a fraction or as a terminating decimal, i.e. they are not rational numbers.

A number such as $\sqrt{2}$ is an irrational number and can only be expressed exactly when left as $\sqrt{2}$. In this form it is called a surd.

Note that $\sqrt{2}$ means the positive square root of 2.

Simplifying surds

Many surds can be simplified.

For example, $\sqrt{18} = \sqrt{9 \times 2} = \sqrt{9} \times \sqrt{2} = 3\sqrt{2}$

And $\sqrt{8} + \sqrt{2} = \sqrt{4 \times 2} + \sqrt{2} = 2\sqrt{2} + \sqrt{2} = 3\sqrt{2}$

In both cases, $3\sqrt{2}$ is the simplest possible surd form.

> **When a calculation involves surds, you should give your answer in the simplest possible surd form.**

Operations on surds

An expression such as $(3 - \sqrt{2})(2 - \sqrt{3})$ can be expanded,

i.e. $(3 - \sqrt{2})(2 - \sqrt{3}) = 6 - 3\sqrt{3} - 2\sqrt{2} + \sqrt{6}$ $(\sqrt{2} \times \sqrt{3} = \sqrt{2 \times 3})$

When the same surd occurs in each bracket the expansion can be simplified.

For example,

$(5 - 2\sqrt{3})(3 + 2\sqrt{3}) = 15 - 6\sqrt{3} + 10\sqrt{3} - 12$

$$(2\sqrt{3} \times 2\sqrt{3} = 4\sqrt{9} = 12)$$

$$= 3 + 4\sqrt{3}$$

In particular, expressions of the form $(a - \sqrt{b})(a + \sqrt{b})$ simplify to a single rational number.

For example,

$(5 - 2\sqrt{3})(5 + 2\sqrt{3}) = 5^2 - (2\sqrt{3})^2$

$$((2\sqrt{3})^2 = 4\sqrt{3}\sqrt{3} = 4\sqrt{9} = 4 \times 3)$$

$$= 25 - 12 = 13$$

Example

Simplify $(2 - \sqrt{5})(3 + 2\sqrt{5})$

$(2 - \sqrt{5})(3 + 2\sqrt{5}) = 6 - 3\sqrt{5} + 4\sqrt{5} - 10$

$$(2\sqrt{5} \times \sqrt{5} = 2\sqrt{25} = 10)$$

$$= -4 + \sqrt{5}$$

Rationalising the denominator

When a fraction has a surd in the denominator, it can be transferred to the numerator.

When the denominator is a single surd, multiplying the fraction, top and bottom, by that surd will change the denominator into a rational number.

For example,

$$\frac{2 + \sqrt{3}}{\sqrt{5}} = \frac{2 + \sqrt{3}}{\sqrt{5}} \times \frac{\sqrt{5}}{\sqrt{5}}$$

$$= \frac{2\sqrt{5} + \sqrt{15}}{5}$$

When the denominator is of the form $a + \sqrt{b}$, multiplying the fraction, top and bottom, by $a - \sqrt{b}$ will change the denominator into a rational number.

For a denominator of the form $a - \sqrt{b}$ multiply top and bottom by $a + \sqrt{b}$

Example

Rationalise the denominator and simplify $\dfrac{\sqrt{2} - 1}{\sqrt{3}(\sqrt{2} + 3)}$

This fraction has a single surd and a bracket in the denominator. Do not attempt to rationalise them both at the same time. We will start with rationalising the single surd.

$$\frac{\sqrt{2} - 1}{\sqrt{3}(\sqrt{2} + 3)} = \frac{\sqrt{3}(\sqrt{2} - 1)}{\sqrt{3} \times \sqrt{3}(\sqrt{2} + 3)} = \frac{\sqrt{6} - \sqrt{3}}{3(\sqrt{2} + 3)}$$

$$= \frac{\sqrt{6} - \sqrt{3}}{3(\sqrt{2} + 3)} \times \frac{\sqrt{2} - 3}{\sqrt{2} - 3} = \frac{\sqrt{12} - \sqrt{6} - 3\sqrt{6} + 3\sqrt{3}}{3(\sqrt{2}^2 - 9)}$$

$$= \frac{2\sqrt{3} - 4\sqrt{6} + 3\sqrt{3}}{3(2 - 9)} = \frac{4\sqrt{6} - 5\sqrt{3}}{21}$$

We have written down every step in this example, but you should be able to do some of these steps in your head.

Exercise 1.3

Expand and simplify when possible.

1 $(3 - 2\sqrt{3})(\sqrt{3} - \sqrt{2})$ **2** $(\sqrt{2} - \sqrt{5})^2$ **3** $(1 - (\sqrt{3} + \sqrt{2}))^2$

Rationalise the denominator of each surd and simplify when possible.

4 $\dfrac{2}{\sqrt{2}}$

7 $\dfrac{1 - \sqrt{2}}{1 + \sqrt{2}}$

5 $\dfrac{2\sqrt{2}}{\sqrt{3}}$

8 $\dfrac{\sqrt{3}}{2\sqrt{3} - 5\sqrt{5}}$

6 $\dfrac{1}{3 - \sqrt{2}}$

9 $\dfrac{\sqrt{8}}{\sqrt{2}(\sqrt{3} - \sqrt{2})}$

Propositions

A sentence such as 'Sonia went to school today' is a closed sentence, but 'She went to school today' is not closed because it contains the variable 'she', who could be any female.

Closed sentences are called statements or **propositions** and are denoted by p, q, etc.

A proposition is either true or false.

Negation

The proposition 'It is not raining' contradicts the proposition 'It is raining'.

'It is not raining' is called the **negation** of 'It is raining'.

If p is the proposition 'It is raining', the negation of p is denoted by $\sim p$.

Truth tables

For the proposition p: 'It is raining', if p is true then $\sim p$ is false.

But if p is false, then $\sim p$ is true.

We can show this logic in table form (called a truth table).
We use 1 to represent true and 0 to represent false.
The numbers in each column are called the **truth values**.

p	$\sim p$
1	0
0	1

Conjunction

The statements p: 'It is raining' and q: 'It is cold' can be combined as 'It is raining *and* it is cold'. This is called a **conjunction** of two propositions.

Using the symbol $\wedge$ to mean 'and' we write this conjunction as $p \wedge q$

We can construct a truth table for $p \wedge q$

p can be true or false, q can also be true or false. We put all possible combinations of 1 (true) and 0 (false) for p and q in the first two columns. Then, reading across, we can complete the third column for $p \wedge q$ (If either p or q is false, then p and q must be false.)

p	q	$p \wedge q$
1	1	1
1	0	0
0	1	0
0	0	0

Disjunction

The statements p: 'It is raining' and q: 'It is cold' can be combined as 'It is raining *and/or* 'it is cold'. This is called a **disjunction** of two propositions and the word 'and' is implied so it would normally be written as 'It is raining *or* it is cold.'

Using the symbol $\vee$ to mean 'or' we write this disjunction as $p \vee q$

We can construct a truth table for $p \vee q$

p can be true or false, q can also be true or false. As before, we put all possible combinations of 1 and 0 for p and q in the first two columns. Then reading across we can complete the third column for $p \vee q$. (If either p or q or both are true, then p or q must be true.)

p	q	$p \vee q$
1	1	1
1	0	1
0	1	1
0	0	0

Conditional statements

If 'it is raining' then 'it is cold' is called a **conditional statement**.

Using the symbol $\rightarrow$ to mean 'If ... then ...' we write $p \rightarrow q$

The proposition p is called the **hypothesis** and the proposition q is called the **conclusion**.

> **In logic, $p \rightarrow q$ is true except when a true hypothesis leads to a false conclusion.**

For example, if p is '5 is a prime number' and q is '6 is a prime number' then in logic $p \rightarrow q$ is false.

The truth table for $p \rightarrow q$ is such that $p \rightarrow q$ is false for only one combination of p and q: p true and q false.

p	q	$p \rightarrow q$
1	1	1
1	0	0
0	1	1
0	0	1

The **converse** of $p \rightarrow q$ is $q \rightarrow p$

For example, the converse of '5 is a prime number' $\rightarrow$ '6 is a prime number' is '6 is a prime number' $\rightarrow$ '5 is a prime number'.

Also the converse of 'It is cold' $\rightarrow$ 'It is raining' is 'It is raining' $\rightarrow$ 'It is cold'.

The **inverse** of $p \rightarrow q$ is $\sim p \rightarrow \sim q$

For example the inverse of '5 is a prime number' $\rightarrow$ '6 is a prime number' is '5 is not a prime number' $\rightarrow$ '6 is not a prime number'.

and the inverse of 'It is raining' $\rightarrow$ 'It is cold' is 'It is not raining' $\rightarrow$ 'It is not cold'.

The **contrapositive** of $p \rightarrow q$ is $\sim q \rightarrow \sim p$

For example the contrapositive of '5 is a prime number' $\rightarrow$ '6 is a prime number' is '6 is not a prime number' $\rightarrow$ '5 is not a prime number'.

Bi-conditional statements

A **bi-conditional statement** is the conjunction of the conditional statement $p \rightarrow q$ with its converse $q \rightarrow p$, that is $(p \rightarrow q) \wedge (q \rightarrow p)$. This reads 'if p then q and if q then p'.

For example, 'If it is raining then it is cold' and 'If it is cold then it is raining' is a bi-conditional statement.

'If it is raining then it is cold' and 'If it is cold then it is raining' can be written simply as 'It is raining' if and only if 'It is cold'.'

Using the symbol $\Leftrightarrow$ to mean 'if and only if' we can write 'It is raining' $\Leftrightarrow$ 'It is cold' and $(p \rightarrow q) \wedge (q \rightarrow p)$ can be written as $p \Leftrightarrow q$

We can construct a truth table for $p \Leftrightarrow q$

Start with the truth table for $p \rightarrow q$, then add a column for $q \rightarrow p$. Lastly, add a column for the conjunction of the third and fourth columns.

p	q	$p \rightarrow q$	$q \rightarrow p$	$(p \rightarrow q) \wedge (q \rightarrow p)$
1	1	1	1	1
1	0	0	1	0
0	1	1	0	0
0	0	1	1	1

This table can now be written as a simpler truth table for a bi-conditional statement.

p	q	$p \Leftrightarrow q$
1	1	1
1	0	0
0	1	0
0	0	1

The table shows that $p \Leftrightarrow q$ is true only when p and q are both true or both false.

Compound statements

A *compound statement* combines two or more propositions using a combination of two or more of the symbols $\sim$, $\wedge$, $\vee$, $\rightarrow$, $\leftarrow$.

A bi-conditional statement, $(p \rightarrow q) \wedge (q \rightarrow p)$, is an example of a compound statement.

Example

Let p, q and r be the propositions:

p: 'Students play soccer', q: 'Students play cricket', r: 'Students play basketball'.

Express the compound statement 'Students play soccer or basketball but not both and students play cricket' in symbolic form.

'Students play soccer or basketball' is $p \vee r$. 'Students do not play both soccer and basketball' is $\sim(p \wedge r)$.

'Students play soccer or basketball but not both' is $(p \vee r) \wedge \sim(p \wedge r)$.

Adding 'and students play cricket' to this gives $(p \vee r) \wedge \sim(p \wedge r) \wedge q$.

The truth table for a compound statement can be constructed in a similar way to the bi-conditional table above.

Example

Construct a truth table for the compound statement $p \vee (\sim q \wedge p)$

p	q	$\sim q$	$\sim q \wedge p$	$p \vee (\sim q \wedge p)$
1	1	0	0	1
1	0	1	1	1
0	1	0	0	0
0	0	1	0	0

Always start with p and q. Then add columns in stages to build up the compound statement.

Equivalence

Two statements are logically equivalent when their truth values are the same, that is in the completed truth tables the final columns are identical.

Example

Determine whether the statements $p \wedge q$ and $\sim p \to q$ are logically equivalent.

We construct a truth table for each statement:

p	q	$\sim p$	$p \wedge q$	$\sim p \to q$
1	1	0	1	1
1	0	0	0	1
0	1	1	0	1
0	0	1	0	0

The truth values for $p \wedge q$ and $\sim p \to q$ are not the same. Therefore the statements are logically not equivalent. We write $p \wedge q \neq \sim p \to q$

Identity law

This law states that $p \wedge p$ and $p \vee p$ are both equivalent to p.

Algebra of propositions

The symbols $\wedge$ and $\vee$ are called logical connectors.

These connectors are commutative, that is $p \wedge q = q \wedge p$ and $p \vee q = q \vee p$

They are also associative, that is $(p \wedge q) \wedge r = p \wedge (q \wedge r)$

They are also distributive over each other and over the conditional $\to$, for example

$p \wedge (q \vee r) = (p \wedge q) \vee (p \wedge r)$ and $p \vee (q \wedge r) = (p \vee q) \wedge (p \vee r)$

$p \wedge (q \to r) = (p \wedge q) \to (p \wedge r)$ and $p \vee (q \to r) = (p \vee q) \to (p \vee r)$

These properties can be proved using truth tables.

The properties can also be used to prove the equivalence between two compound statements.

It can also be shown that $p \to q = \sim q \to \sim p$

Example

Use algebra to show that
$p \wedge (p \vee q) = p \vee (p \wedge q)$

$p \wedge (p \vee q) = (p \wedge p) \vee (p \wedge q)$
   using the distributive law

$= p \vee (p \wedge q)$
   using the distributive law

Exercise 1.4

In this exercise, p, q and r are propositions.

1   Write down the contrapositive of $\sim p \wedge q$

2   (a) Construct a truth table for $p \to \sim q$ and $p \vee \sim q$
    (b) State, with a reason, whether $p \to \sim q$ and $p \vee \sim q$ are logically equivalent.

3   p: 'It is raining', q: 'It is cold', r: 'The sun is shining'.
    Using logic symbols, write down in terms of p, q and r the statement:
    'The sun is shining and it is cold and it is not raining'.

1.5 Direct proof

- To construct simple proofs, specifically direct proofs
- Proof by the use of counter examples

You need to know

- The basic rules of logic
- How to solve a quadratic equation by factorisation or by the formula
- How to find the area of a triangle

Direct proof

Mathematics is the study of numbers, shapes, space and change. Mathematicians look for patterns and formulate conjectures. They then try to prove the truth, or otherwise, of conjectures by proof that is built up from axioms. The axioms are the basic rules or definitions, and all other facts can be derived from these by deduction, that is by using true inferences from those rules (an inference is the same as the logic conditional $\rightarrow$). (We can use the game of chess as an analogy – the basic rules are the moves that are allowed for each piece, and games are built up from these moves.)

For example, x^n is defined to mean n lots of x multiplied together, i.e. $x \times x \times x \times \ldots \times x$, and from this definition we can deduce that $x^a \times x^b = x^{a+b}$

Example

Prove that if $4(x - 5) = 8$ then $x = 7$

Starting with $4(x - 5) = 8$

$\Rightarrow \qquad 4x - 20 = 8$ Using the distributive law

$\Rightarrow \qquad\qquad 4x = 28$ Adding 20 to each side keeps the equality true

$\Rightarrow \qquad\qquad\quad x = 7$ Dividing each side by 4 keeps the equality true

$\therefore \qquad 4(x - 5) = 8 \;\Rightarrow\; x = 7$

This is an example of direct proof by deduction, i.e. to prove $p \Rightarrow q$, start with p then deduce $p \Rightarrow r \Rightarrow s \Rightarrow q$, so $p \Rightarrow q$

(Note that this is a proof of an *implication* $p \Rightarrow q$. We know from Topic 1.4 that whether p is true or q is true is another question.)

We know from logic that if $p \rightarrow q$ is true then the contrapositive $\sim q \rightarrow \sim p$ is also true. Therefore we can also say that if $x \neq 7$ then $4(x - 5) \neq 8$

The converse of $4(x - 5) = 8 \Rightarrow x = 7$, i.e. $x = 7 \Rightarrow 4(x - 5) = 8$ is also true in this case, but *the converse of a true implication is not always true*.

For example, 'A polygon is a square $\Rightarrow$ a polygon has four equal sides' is true

but 'A polygon has four equal sides $\Rightarrow$ a polygon is a square' is not true because a rhombus has four equal sides.

Therefore the converse of an implication needs to be proved to be true.

Example

Prove that the sum of the interior angles of any triangle is $180°$.

ABC is any triangle. DE is parallel to AB.

$\angle DCA = \angle CAB$ Alternate angles are equal

$\angle ECB = \angle CBA$ Alternate angles are equal

$\angle DCA + \angle ACB + \angle ECB = 180°$ Supplementary angles

$\Rightarrow \quad \angle CAB + \angle ACB + \angle ECB = 180°$

$\Rightarrow \quad$ the sum of the interior angles of any triangle is $180°$.

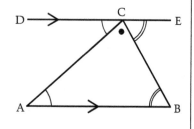

Use of counter examples

As well as it being necessary to prove that a statement is true, it is also important to prove that a statement is false. This is particularly important for the converse of a true implication.

A statement can be shown to be false if we can find just one example that disproves it. This is called a ***counter example***.

For example, $a > 0 \Rightarrow a^2 > 0$ is true, but the converse $a^2 > 0 \Rightarrow a > 0$ is false.

We can use $a^2 = 9 \Rightarrow a = 3$ or -3 as a counter example because $-3 \not> 0$.

For example, the statement 'all prime numbers are odd' is not true. We can prove this using the counter example '2 is a prime number and 2 is not an odd number'.

Example

Use a counter example to prove that the converse of the true statement: 'n is an integer' $\Rightarrow$ 'n^2 is an integer' is false.

The converse of the given statement is
'n^2 is an integer' $\Rightarrow$ 'n is an integer'.

$(\sqrt{2})^2 = 2$ is an integer but $\sqrt{2}$ is not an integer.

Therefore 'n^2 is an integer' $\Rightarrow$ 'n is an integer' is false.

Exercise 1.5

1 Prove that if $x^2 - 3x + 2 = 0$ then $x = 1$ or $x = 2$

2 Find a counter example to show that $a > b \Rightarrow a^2 > b^2$ is not true.

3 **(a)** Prove that 'n is an odd integer $\Rightarrow n^2$ is an odd integer'.
   (Start with $n = 2k + 1$ where k is any integer.)

   (b) Use a counter example to show that the converse of the statement in **(a)** is false.

4 **(a)** Prove that '$x^2 + bx + c = 0$ has equal roots $\Rightarrow b^2 = 4c$'

   (b) Prove that the converse of the statement in **(a)** is also true.

5 In the diagram, D is the midpoint of AB.
   Prove that the area of triangle ABC
   is twice the area of triangle ADC.

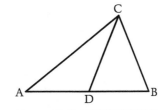

Learning outcomes

- Establish simple proofs by using the principle of mathematical induction

You need to know

- The set of positive integers is denoted by $\mathbb{N}$

- Any even number can be written as $2k$ and any odd number can be written as $2k - 1$ for $k \in \mathbb{N}$

- A natural number is a member of the set 1, 2, 3, 4, ...

Proof by induction

Consider these results: $1 + 1^2 = 2$, $2 + 2^2 = 6$, $3 + 3^2 = 12$, $4 + 4^2 = 20$

In every case, the right-hand side is a multiple of 2.

This suggests that the proposition
'for any positive integer n, $n + n^2$ is a multiple of 2'
is true but it does not prove it.

We can prove it using a method called **_mathematical induction_**.

We call the proposition $p(n)$ and rephrase it as '$n + n^2 = 2m$ for n, $m \in \mathbb{N}$'.

We start with the proposition, that
when $n = k$, '$k + k^2$ is a multiple of 2, $k \in \mathbb{N}$'.   [1]

The next step is to replace k by $k + 1$ (i.e. by the next consecutive integer)

$$\Rightarrow \quad (k + 1) + (k + 1)^2$$
$$= k + 1 + k^2 + 2k + 1$$
$$= (k + k^2) + 2(k + 1) \text{ which is also a multiple of 2.}$$

From [1] this is a multiple of 2

Therefore we have shown that *if* for any integer, n, $n + n^2$ is a multiple of 2

then $(n + 1) + (n + 1)^2$ is also a multiple of 2.   [2]

We know that $p(1)$ is true, i.e. $1 + 1^2$ is a multiple of 2.

Therefore   [2] shows that $p(1) \Rightarrow p(2)$ so $p(2)$ is true,

then again   [2] shows that $p(2) \Rightarrow p(3)$ so $p(3)$ is true,

again   [2] shows that $p(3) \Rightarrow p(4)$ so $p(4)$ is true, ...

This process can be continued indefinitely, i.e. for all positive integers.

Therefore we have proved that, for any positive integer n, '$n + n^2$ is a multiple of 2'.

An analogy for proof by induction is a row of evenly spaced dominoes standing on end.

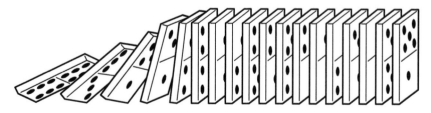

If you can show that pushing over any domino will make the next domino fall, then pushing over the first domino will make the whole row fall, one domino after the other.

Proof by induction can be used to prove many results that are generalised to cover any positive integer from a result proved for a particular integer.

The proof has three distinct steps:

1   Let $p(n)$ be a proposition involving n, then assume that $p(k)$ is true. Prove directly that $p(k+1)$ is true. (Note that k is an arbitrary and convenient integer.)

2   Prove that $p(1)$ is true.

3   Combine steps **1** and **2** to prove that $p(2)$, $p(3)$, $p(4)$, ... are true.

Example

Prove by induction that $10^n - 1$ is a multiple of 9, $n \in \mathbb{N}$.

Assume that $p(k)$ is $10^k - 1 = 9m$, where k and m are natural numbers.

Replace k by $k + 1$, giving $10^{k+1} - 1 = 10 \times 10^k - 1$

$$= 10 \times 10^k - 10 + 9$$
$$= 10(10^k - 1) + 9 = 10(9m) + 9 \text{ which is a multiple of } 9.$$

Therefore when $10^k - 1$ is a multiple of 9 then $10^{k+1} - 1$ is also a multiple of 9.   [1]

$p(1)$ is the proposition $10^1 - 1$ is a multiple of 9, and $10^1 - 1 = 9$, therefore from [1] $10^2 - 1$ is a multiple of 9, from [1] again $10^3 - 1$ is a multiple of 9, ...

Therefore $10^n - 1$ is a multiple of 9, for all $n \in \mathbb{N}$.

Example

Prove by induction that the sum of the first n odd numbers is n^2.

(The second odd number is $2(2) - 1$, the third odd number is $2(3) - 1$, ... the nth odd number is $2n - 1$)

Let $p(n)$ be the proposition that $1 + 3 + 5 + ... + (2n - 1) = n^2$

Assume that $1 + 3 + 5 + ... + (2k - 1) = k^2$

Then adding the next odd number to both sides gives

$$1 + 3 + 5 + ... + (2k - 1) + (2k + 1) = k^2 + 2k + 1 = (k + 1)^2$$

Therefore when $1 + 3 + 5 + ... + (2k - 1) = k^2$ then $1 + 3 + 5 + ... + (2k + 1) = (k + 1)^2$

i.e. when the sum of the first k odd numbers is k^2, the sum of the first $(k + 1)$ odd numbers is $(k + 1)^2$

Now $p(1)$: $1 = 1^2$, so the sum of the first two odd numbers is 2^2, and so on.

Therefore the sum of the first n odd numbers is n^2.

Exercise 1.6

Prove by induction that:

1 $n^3 - n$ is a multiple of 6 for all positive integral values of n

2   the sum of the first n natural numbers is $\frac{1}{2}n(n + 1)$

Learning outcomes

- To apply the remainder theorem
- To use the factor theorem to find factors and to evaluate unknown coefficients

You need to know

- The meaning of the notation $f(a)$ for the function $f(x)$ where a is a value of x
- How to expand expressions of the form $(ax + c)$(a polynomial)
- How to factorise a quadratic expression

Polynomials

The general form of a polynomial expression is

$$a_nx^n + a_{n-1}x^{n-1} + \;.... + a_2x^2 + a_1x + a_0$$

where n is a positive integer, $a_n, a_{n-1}, ... a_1, a_0$ are real numbers and $a_n \neq 0$

The *order of a polynomial* is the highest power of x. For example, $x^5 - x^2 + 2$ has order 5.

Identical polynomials

Two polynomials are identical when they have the same order and when each power of x has equal coefficients.

For example, $x^4 - 5x^2 + 2 \equiv ax^5 + bx^4 + cx^2 + dx + e$ if and only if $a = 0$ (the order must be the same) and $b = 1, c = -5, d = 0, e = 2$ (coefficients must be equal).

The remainder theorem

When 17 is divided by 3 the result can be written as 5, remainder 2, i.e. $\frac{17}{3} = 5 + \frac{2}{3}$ which can be written as $17 = 5 \times 3 + 2$. In this form, 5 is called the quotient.

When $f(x) = x^3 - 7x^2 + 6x - 2$ is divided by $x - 2$, we get a quotient and a remainder. The relationship between these quantities can be written as

$$f(x) = x^3 - 7x^2 + 6 \equiv \text{(quotient)}(x - 2) + \text{remainder}$$

Substituting 2 for x eliminates the term containing the quotient, giving $f(2) = \text{remainder}$.

Now $f(2) = 2^3 - 7(2^2) + 6 = -14$, so when $f(x) = x^3 - 7x^2 + 6x - 2$ is divided by $x - 2$, the remainder is -14.

This is an illustration of the general case:
when a polynomial $f(x)$ is divided by $(ax - b)$ then

$$f(x) \equiv \text{(quotient)}(ax - b) + \text{remainder} \Rightarrow f\left(\frac{b}{a}\right) = \text{remainder}$$

This result is called the *remainder theorem* and can be summarised as: when a polynomial $f(x)$ is divided by $(ax - b)$, the remainder is $f\left(\frac{b}{a}\right)$.

Example

Find the remainder when $2x^3 + 7x^2 - 3$ is divided by $2x - 1$.

Let $f(x) = 2x^3 + 7x^2 - 3$

$2x - 1 = 0$ when $x = \frac{1}{2}$, therefore when $2x^3 + 7x^2 - 3$ is divided by $2x - 1$ the remainder is $f(\frac{1}{2})$.

$f(\frac{1}{2}) = 2(\frac{1}{2})^3 + 7(\frac{1}{2})^2 - 3 = 1$

$\therefore$ the remainder is 1.

Example

When $x^3 - ax^2 + x + b$ is divided by $x - 1$, the remainder is 4 and when $x^3 - ax^2 + x + b$ is divided by $x - 3$, the remainder is 16. Find the values of a and b.

Using the remainder theorem gives
$1 - a + 1 + b = 4$ and $27 - 9a + 3 + b = 16$,

i.e. $b - a = 2$ [1]

and $b - 9a = -14$ [2]

$[1] - [2] \Rightarrow 8a = 16 \Rightarrow a = 2$

Substituting 2 for a in $[1] \Rightarrow b = 4$

The factor theorem

When $(x - a)$ is a factor of the polynomial $f(x)$,
the remainder is zero $\Rightarrow f(a) = 0$

**This is the factor theorem,
i.e. if, for a polynomial $f(x)$, $f(a) = 0$ then $x - a$ is a factor of $f(x)$.**

For example, when $x = 3$,
$x^4 - 3x^3 - 3x^2 + 11x - 6 = 81 - 81 - 27 + 33 - 6 = 0$
Therefore $x - 3$ is a factor of $x^4 - 3x^3 - 3x^2 + 11x - 6$

Example

Given that $ax^3 + 3x^2 - b$ has a factor $2x - 1$ and leaves a remainder
-5 when divided by $x + 2$, find the values of a and b.

$\frac{a}{8} + \frac{3}{4} - b = 0 \Rightarrow a - 8b = -6$ [1] Using the factor theorem with $x = \frac{1}{2}$

$-8a + 12 - b = -5 \Rightarrow -8a - b = -17$ [2]
Using the remainder theorem with $x = -2$

$8 \times [1] + [2] \Rightarrow -65b = -65 \Rightarrow b = 1$

Substituting 1 for b in [1] gives $a = 2$

The factor theorem can be used to find factors of polynomials.

Example

Factorise $x^3 - x^2 - x - 2$.

If $x - \alpha$ is a factor of $x^3 - x^2 - x - 2$ then
$x^3 - x^2 - x - 2 \equiv (x - \alpha)(x^2 + bx + c)$

$\therefore - \alpha c = -2$, so possible values of α are ± 1 and ± 2.

Try $\alpha = 1$: $(1)^3 - (1)^2 - (1) - 2 \neq 0$ so $x - 1$ is not a factor.

Try $\alpha = -1$: $(-1)^3 - (-1)^2 - (-1) - 2 \neq 0$ so $x + 1$ is not a factor.

Try $\alpha = 2$: $(2)^3 - (2)^2 - (2) - 2 = 0$ therefore $x - 2$ is a factor.

$\therefore x^3 - x^2 - x - 2 \equiv (x - 2)(x^2 + bx + c) \equiv x^3 + (b - 2)x^2 + (c - 2b)x - 2c$

Comparing coefficients of x^2 and the constant gives $-1 = b - 2$
so $b = 1$ and $-2 = -2c$ so $c = 1$

$\therefore x^3 - x^2 - x - 2 \equiv (x - 2)(x^2 + x + 1)$

Exercise 1.7

1 Given that $(x - 1)$ and $(x + 2)$ are factors of $x^3 + ax^2 + bx - 6$, find
the values of a and b.

2 $f(x) = 5x^3 - px^2 + x - q$. When $f(x)$ is divided by $x - 2$, the
remainder is 3.

Given that $(x - 1)$ is a factor of $f(x)$, find p and q.

Factors of $a^2 - b^2$

$a^2 - b^2$ is the difference between two squares, so $a^2 - b^2 \equiv (a - b)(a + b)$

Factors of $a^3 - b^3$

From the factor theorem, when $a = b$, $a^3 - b^3 = b^3 - b^3 = 0$

Therefore $a - b$ is a factor of $a^3 - b^3 \Rightarrow a^3 - b^3 \equiv (a - b)(a^2 + ab + b^2)$

(You can verify this by expanding the right-hand side.)

$(a^2 + ab + b^2)$ cannot be factorised.

$$\textbf{Therefore } a^3 - b^3 \equiv (a - b)(a^2 + ab + b^2)$$

For example, $x^3 - 8 = x^3 - 2^3 = (x - 2)(x^2 + 2x + 2)$

Factors of $a^4 - b^4$

$a^4 - b^4$ is the difference between two squares,

i.e. $(a^2)^2 - (b^2)^2 = (a^2 - b^2)(a^2 + b^2)$

$$= (a - b)(a + b)(a^2 + b^2)$$

using the factors of the difference between two squares twice.

$(a^2 + b^2)$ cannot be factorised.

$$\textbf{Therefore } a^4 - b^4 \equiv (a - b)(a + b)(a^2 + b^2)$$

For example, $x^4 - 16 = x^4 - 2^4 = (x - 2)(x + 2)(x^2 + 4)$

Factors of $a^5 - b^5$

From the factor theorem, when $a = b$, $a^5 - b^5 = b^5 - b^5 = 0$

Therefore $a - b$ is a factor of $a^5 - b^5$
$\Rightarrow a^5 - b^5 \equiv (a - b)(a^4 + a^3b + a^2b^2 + ab^3 + b^4)$

(You can verify this by expanding the right-hand side.)

$(a^4 + a^3b + a^2b^2 + ab^3 + b^4)$ has no linear factors.

$$\textbf{Therefore } a^5 - b^5 \equiv (a - b)(a^4 + a^3b + a^2b^2 + ab^3 + b^4)$$

For example, $x^5 - 32 = x^5 - 2^5 = (x - 2)(x^4 + 2x^3 + 4x^2 + 8x + 16)$

Factors of $a^6 - b^6$

$a^6 - b^6 = (a^3)^2 - (b^3)^2$, which is the difference between two squares.

Therefore $a^6 - b^6 = (a^3 - b^3)(a^3 + b^3)$

$$= (a - b)(a^2 + ab + b^2)(a^3 + b^3)$$

Now when $a = -b$, $a^3 + b^3 = -b^3 + b^3 = 0$

Therefore $a + b$ is a factor of $a^3 + b^3 \Rightarrow a^3 + b^3 = (a + b)(a^2 - ab + b^2)$

(You can verify this by expanding the right-hand side.)

Neither $(a^2 + ab + b^2)$ nor $(a^2 - ab + b^2)$ can be factorised.

Therefore $a^6 - b^6 \equiv (a - b)(a + b)(a^2 + ab + b^2)(a^2 - ab + b^2)$

For example, $x^6 - 64 = x^6 - 2^6 = (x - 2)(x + 2)(x^2 + 2x + 4)(x^2 - 2x + 4)$

These results can be used to factorise any polynomial that can be expressed in one of the forms given above.

Example

Factorise $8x^3 - 27$ completely.

$8x^3 - 27$ can be written as $(2x)^3 - (3)^3$.

Using $\quad a^3 - b^3 \equiv (a - b)(a^2 + ab + b^2)$

$\qquad$ and replacing a by $2x$ and b by 3 gives

$\qquad 8x^3 - 27 = (2x - 3)\{(2x)^2 + (2x)(3) + (3)^2\}$

$\qquad\qquad\qquad = (2x - 3)(4x^2 + 6x + 9)$

Example

Show that $(x + 1)^4 - x^4 \equiv (x + 1)^3 + x(x + 1)^3 + x^2(x + 1) + x^3$

Using $\quad a^4 - b^4 \equiv (a - b)(a + b)(a^2 + b^2)$

$\qquad$ and expanding the last two bracket gives

$\qquad a^4 - b^4 \equiv (a - b)(a^3 + a^2b + ab^2 + b^3)$

Replacing a by $x + 1$ and b by x gives

$(x + 1)^4 - x^4 \equiv (x + 1 - x)\{(x + 1)^3 + x(x + 1)^2 + x^2(x + 1) + x^3\}$

$\qquad\qquad\quad \equiv (x + 1)^3 + x(x + 1)^2 + x^2(x + 1) + x^3$

Example

(a) Show that $7x^3 - 3x^2 - 3x - 1 \equiv 8x^3 - (x + 1)^3$

(b) Hence or otherwise factorise $7x^3 - 3x^2 - 3x - 1$

(a) $7x^3 - 3x^2 - 3x - 1 \equiv 8x^3 - x^3 - 3x^2 - 3x - 1$

$\qquad\qquad\qquad\qquad\quad \equiv 8x^3 - (x^3 + 3x^2 + 3x + 1)$

$\qquad\qquad\qquad\qquad\quad \equiv 8x^3 - (x + 1)^3$

(b) $7x^3 - 3x^2 - 3x - 1 \equiv 8x^3 - (x + 1)^3$

$\qquad$ Using $a^3 - b^3 \equiv (a - b)(a^2 + ab + b^2)$

$\qquad$ and replacing a by $2x$ and b by $(x + 1)$ gives

$\qquad 8x^3 - (x + 1)^3 = \{2x - (x + 1)\}\{(4x^2 + 2x(x + 1) + (x + 1)^2\}$

$\qquad\qquad\qquad\qquad = (x - 1)(7x^2 + 4x + 1)$

$\qquad \therefore \ 7x^3 - 3x^2 - 3x - 1 \equiv (x - 1)(7x^2 + 4x + 1)$

Exercise 1.8

---

1 Factorise $8x^3 + 1$ completely. (Hint: $(-1)^3 = -1$)

2 Show that $x^4 - (x - 2)^4 = 8(x - 1)(x^2 - 2x + 2)$

1.9   Quadratic and cubic equations

Learning outcomes

- To investigate the nature of the roots of a quadratic equation and the relationship between the sum and product of these roots and the coefficients of $ax^2 + bx + c = 0$

- To use the relationship between the sum of the roots, the product of the roots, the sum of the product of the roots pairwise and the coefficients of $a^3 + bx^2 + cx + d = 0$

You need to know

- How to expand brackets of the form $(ax + b)$(a polynomial)

Polynomial equations

A polynomial equation has the form

$$a_n x^n + a_{n-1} x^{n-1} + \ldots + a_2 x^2 + a_1 x + a_0 = 0$$

The roots of a polynomial equation are the values of x that satisfy the equation.

The order of the polynomial gives the number of roots of the equation. Some, or all, of these roots may not be real. For example, the quadratic equation $x^2 + x + 2 = 0$ has two roots, although neither of them are real. (You will discover the nature of these roots if you study Pure Mathematics Unit 2.)

The nature of the roots of a quadratic equation

The general form of a quadratic equation is $ax^2 + bx + c = 0$.

The values of x that satisfy this equation are given by

$$x = \frac{-b \pm \sqrt{b^2 - 4ac}}{2a}$$ and it is the value of $b^2 - 4ac$ that determines the nature of these roots.

Note that $b^2 - 4ac$ is called the **discriminant**.

When $b^2 - 4ac > 0$, $\pm\sqrt{b^2 - 4ac}$ has two real and different values, therefore the roots are real and different.

When $b^2 - 4ac = 0$, $\pm\sqrt{b^2 - 4ac} = 0$

$$\therefore \quad x = -\frac{b}{2a} + 0 \text{ and } x = -\frac{b}{2a} - 0$$

so there is only one value of x that satisfies the equation and the equation is said to have a **repeated root**.

When $b^2 - 4ac < 0$, $\pm\sqrt{b^2 - 4ac}$ has no real value, so the equation has no real roots.

The relationship between the coefficients of a quadratic equation and the roots

The general form of a quadratic equation is $ax^2 + bx + c = 0$ [1]

If α and β are the roots of this equation, the equation can be expressed as $(x - \alpha)(x - \beta) = 0$

$$\Rightarrow x^2 - (\alpha + \beta)x + \alpha\beta = 0 \quad [2]$$

[1] and [2] are the identical equation, so we say that

$$x^2 + \frac{b}{a}x + \frac{c}{a} \equiv x^2 - (\alpha + \beta)x + \alpha\beta$$

([1] is divided by a, so that the coefficients of x^2 are equal.)

Comparing coefficients of this identity shows that

$$\alpha + \beta = -\frac{b}{a}$$

and

$$\alpha\beta = \frac{c}{a}$$

i.e. the sum of the roots of the equation $ax^2 + bx + c = 0$ is $-\dfrac{b}{a}$ and the product of the roots is $\dfrac{c}{a}$. This is true whether or not the roots are real.

Example

(a) Determine the nature of the roots of the equation
$3x^2 - 2x + 2 = 0$

(b) If α and β are the roots of the equation $3x^2 - 2x + 2 = 0$, find the equation whose roots are $\dfrac{1}{\alpha}$ and $\dfrac{1}{\beta}$.

(a) $3x^2 - 2x + 2 = 0$ so '$b^2 - 4ac$' $= 4 - 4(6) = -20$

Therefore $3x^2 - 2x + 2 = 0$ has no real roots.

(b) $3x^2 - 2x + 2 = 0$ gives $\alpha + \beta = \dfrac{2}{3}$ and $\alpha\beta = \dfrac{2}{3}$

For the equation whose roots are $\dfrac{1}{\alpha}$ and $\dfrac{1}{\beta}$,

the sum of the roots is $\dfrac{1}{\alpha} + \dfrac{1}{\beta} = \dfrac{\beta + \alpha}{\alpha\beta}$

$$= \dfrac{\frac{2}{3}}{\frac{2}{3}} = 1$$

and the product of the roots is $\dfrac{1}{\alpha} \times \dfrac{1}{\beta} = \dfrac{1}{\alpha\beta}$

$$= \dfrac{1}{\frac{2}{3}} = \dfrac{3}{2}$$

Therefore the required equation is $x^2 - x + \dfrac{3}{2} = 0$,
i.e. $2x^2 - 2x + 3 = 0$

Exercise 1.9a

1 One of the roots of the equation $3x^2 - x + c = 0$ is α and the other root is 2α.

Find the value of c.

2 The roots of the equation $x^2 + 3x + 5 = 0$ are α and β.

Find the equation whose roots are $\alpha + 2$ and $\beta + 2$

Cubic equations

The formula for solving a general quadratic equation was known to the ancient Greeks. However, the search for a general solution for the cubic equation continued until a method was developed during the Renaissance in Italy.

This method does lead to a formula, but it is not at all easy to remember and is difficult to work with. It also relies on working with numbers that are not real, and such numbers are not covered in Unit 1. For these reasons it not included here.

If you are interested in finding this formula, search on the internet for 'general solution of cubic equations'.

Did you know?

It is thought that Girolamo Cardano (1501–1576) was the first to publish a general method of solution for the cubic equation.

The roots of a cubic equation

The general form of a cubic equation is $ax^3 + bx^2 + cx + d = 0$

The order of this equation is three, therefore it has three roots.

If these roots are α, β and γ then the equation can be written as

$$(x - \alpha)(x - \beta)(x - \gamma) = 0$$

By expanding this form of the equation and comparing with the general form, we can get a relationship between the roots of a cubic equation and the coefficients of the general form.

$$(x - \alpha)(x - \beta)(x - \gamma) = (x - \alpha)(x^2 - (\beta + \gamma)x + \beta\gamma)$$

$$= x^3 - (\alpha + \beta + \gamma)x^2 + (\alpha\beta + \alpha\gamma + \beta\gamma)x - \alpha\beta\gamma$$

Dividing the general form of the cubic equation by a gives

$$x^3 - (\alpha + \beta + \gamma)x^2 + (\alpha\beta + \alpha\gamma + \beta\gamma)x - \alpha\beta\gamma \equiv x^3 + \frac{b}{a}x^2 + \frac{c}{a}x + \frac{d}{a}$$

Therefore if α, β and γ are the roots of $ax^3 + bx^2 + cx + d = 0$, then

$$\alpha + \beta + \gamma = -\frac{b}{a}$$

$$\alpha\beta + \alpha\gamma + \beta\gamma = \frac{c}{a}$$

$$\alpha\beta\gamma = -\frac{d}{a}$$

For example, the sum of the roots of the equation $x^3 + 2x^2 + 5x - 7 = 0$ is -2, the product of the roots pair-wise is 5, and the product of the roots is 7.

Example

Two of the roots of the equation $x^3 + px^2 + 2x + q = 0$ are 1 and -2. Find the values of p and q.

If α is the third root, then $-1 + \alpha = -p$ Sum of the roots   [1]

$\qquad\qquad\qquad\qquad\qquad -2 + \alpha - 2\alpha = 2$ Pair-wise product of the roots [2]

$\qquad\qquad\qquad\qquad\qquad\qquad\qquad -2\alpha = q$ Product of the roots   [3]

[2] gives $\alpha = -4$,
therefore from [1], $p = 5$
and from [3], $q = 8$

Example

The equation $ax^3 + bx^2 + cx + d = 0$ has roots $\alpha - p$, α and $\alpha + p$. Find a relationship between a, b, c and d.

The sum of the roots is $(\alpha - p) + \alpha + (\alpha + p) = 3\alpha$

$\therefore \qquad\qquad 3\alpha = -\frac{b}{a} \quad \Rightarrow \quad \alpha = -\frac{b}{3a}$

α is a root of the equation, so $x = \alpha$ satisfies the equation,

i.e. $\quad a \times \left(-\dfrac{b}{3a}\right)^3 + b \times \left(-\dfrac{b}{3a}\right)^2 + c \times \left(-\dfrac{b}{3a}\right) + d = 0$

Multiplying by $27a^3 \Rightarrow ab^3 - 3ab^3 + 9a^2bc - 27a^3d = 0$

$\qquad\qquad\qquad\quad \Rightarrow 2ab^3 - 9a^2bc + 27a^3d = 0$

Example

The roots of the equation $2x^3 - x^2 + 3x - 1 = 0$ are α, β and γ.

Find the equation whose roots are $\dfrac{1}{\alpha}, \dfrac{1}{\beta}$ and $\dfrac{1}{\gamma}$.

From the given equation

$$\alpha + \beta + \gamma = \frac{1}{2}$$

$$\alpha\beta + \alpha\gamma + \beta\gamma = \frac{3}{2}$$

$$\alpha\beta\gamma = \frac{1}{2}$$

For the required equation, the sum of the roots is

$$\frac{1}{\alpha} + \frac{1}{\beta} + \frac{1}{\gamma} = \frac{\beta\gamma + \alpha\gamma + \alpha\beta}{\alpha\beta\gamma}$$

$$= 3$$

The product of the roots pair-wise is

$$\frac{1}{\alpha\beta} + \frac{1}{\beta\gamma} + \frac{1}{\alpha\gamma} = \frac{\alpha + \beta + \gamma}{\alpha\beta\gamma}$$

$$= 1$$

The product of the roots is $\dfrac{1}{\alpha\beta\gamma} = 2$

Therefore the required equation is $x^3 - 3x^2 + x - 2 = 0$

Exercise 1.9b

1 Two of the roots of the equation $2x^3 + px^2 + qx - 1 = 0$ are $\frac{1}{2}$ and -1.
   Find the values of p and q.

2 The roots of the equation $x^3 + 2x^2 - 5x + 1 = 0$ are α, β and γ.
   Find the equation whose roots are $\alpha + 1$, $\beta + 1$ and $\gamma + 1$

3 The roots of the equation $2x^3 + x^2 + px - q = 0$ are α, β and $\alpha + \beta$
   Find a relationship between p and q.

Learning outcomes

- To revise basic techniques for simple curve sketching

You need to know

- How to express $ax^2 + bx + c$ in the form $a(x - p)^2 + q$

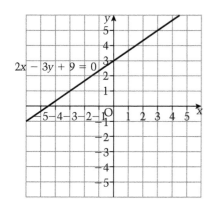

Straight lines

The equation of any straight line can be written as $y = mx + c$ where m is the gradient of the line and c is the intercept on the y-axis.

To sketch the graph of a straight line, you need the coordinates of two points on the line.

The most straightforward points to find are those where the line crosses the axes.

For example, to sketch the line $2x - 3y + 9 = 0$, first find where the line crosses the axes: when $x = 0$, $y = 3$ and when $y = 0$, $x = -4\frac{1}{2}$, so draw the line through $(0, 3)$ and $(-4\frac{1}{2}, 0)$.

Curves

A sketch of a curve should show the shape of the curve and its position on the coordinate axes. It should also show any significant features of the curve such as, for example, where the curve turns. A sketch is not an accurate plot, so these features will in many cases be approximate.

Parabolas

A curve whose equation has the form $y = ax^2 + bx + c$ has a characteristic shape called a parabola.

When $a > 0$, y has a minimum value.

When $a < 0$, y has a maximum value.

In both cases, the curve is symmetrical about the line through the point where the curve turns, as shown in the diagrams.

To sketch the graph of the curve whose equation is $y = ax^2 + bx + c$, you can

either   find the coordinates of the points where the curve crosses the axes (this is easy if $ax^2 + bx + c$ factorises) and then use symmetry to find the coordinates of the turning point

or   express $ax^2 + bx + c$ in the form $a(x - p^2) + q$ to find the coordinates of the turning point together with the fact that the curve crosses the y-axis at the point $(0, c)$.

Example

Sketch the curve whose equation is $y = 2x^2 - 3x + 1$

$$y = 2x^2 - 3x + 1 = (2x - 1)(x - 1)$$

The curve crosses the y-axis at $(0, 1)$ and crosses the x-axis at $(\frac{1}{2}, 0)$ and $(1, 0)$. Therefore the curve is symmetrical about $x = \frac{3}{4}$ (halfway between $x = \frac{1}{2}$ and $x = 1$)

y has a minimum value where $x = \frac{3}{4}$ of $2(\frac{3}{4})^2 - 3(\frac{3}{4}) + 1 = -\frac{1}{8}$

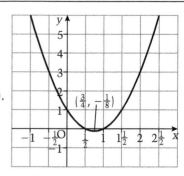

Example

Sketch the curve $y = -x^2 + x - 1$

$y = -(x^2 - x) - 1 = -(x - \frac{1}{2})^2 - \frac{3}{4}$

The curve crosses the y-axis at $(0, -1)$ and y has a maximum value where $x = \frac{1}{2}$ of $-\frac{3}{4}$

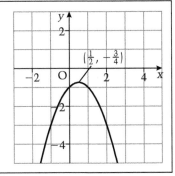

Cubic curves

A curve whose equation is $y = ax^3 + bx^2 + cx + d$ has a characteristic shape.

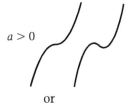 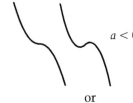

The curve is easy to sketch when the cubic expression factorises.

For example, the graph of $y = (x + 1)(x - 2)(x - 3)$ crosses the x-axis at $(-1, 0)$, $(2, 0)$, $(3, 0)$.

When the brackets are expanded, and comparing with $ax^3 + bx^2 + cx + d$ shows that $a = 1$ and $d = 6$, so the curve crosses the y-axis at $(0, 6)$.

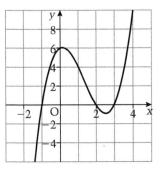

The curve whose equation is $y = \dfrac{1}{x}$

We know that $\frac{1}{0}$ is meaningless, so there is no point on the curve where $x = 0$. When $x > 0$, $y > 0$ and when $x < 0$, $y < 0$ so the curve exists only in the first and third quadrants.

We also know that for positive values, as x gets larger, $\frac{1}{x}$ gets smaller, i.e. as x increases the curve gets closer to the x-axis. Using similar reasoning, as x approaches zero from positive values, y increases.

For negative values of x, as x approaches zero, y decreases, and as x approaches $-\infty$, y increases. The curve gets closer and closer to the axes but never crosses them.

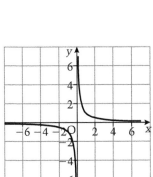

Any line that a curve gets closer and closer to but never crosses is called an **asymptote**.

$y = 0$ and $x = 0$ are asymptotes to the curve $y = \frac{1}{x}$
The curve is symmetric about the line $y = x$

Exercise 1.10

1 Draw sketches of the graphs whose equations are given.

   Mark all significant points on the curves.

   (a) $y = x^2 - 5x + 6$ **(c)** $y = x(x - 1)(x - 3)$
   (b) $y = 3x^2 - x + 1$

2 On the same set of axes, draw sketches of the graphs whose equations are

   $y = \dfrac{2}{x}$ and $2y - 3x + 6 = 0$

Learning outcomes

- To understand how curves are transformed and to use this knowledge to sketch curves

You need to know

- The meaning of translation and reflection
- How to sketch graphs of simple equations

Translations

Consider the curve whose equation is $y = f(x)$ and the curve whose equation is $y = f(x) + 2$

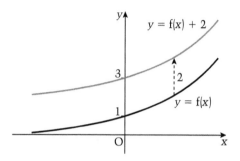

Comparing $y = f(x)$ with $y = f(x) + 2$, we see that for a particular value of x, the value of $f(x) + 2$ is 2 units greater than the value of $f(x)$.

Therefore for equal values of x, points on $y = f(x) + 2$ are 2 units above points on $y = f(x)$, i.e. the curve $y = f(x) + 2$ is a translation of the curve $y = f(x)$ by 2 units in the positive direction of the y-axis.

> **For any function f, the curve whose equation is $y = f(x) + c$ is the translation of the curve whose equation is $y = f(x)$ by c units parallel to the y-axis.**

Now consider the curve whose equation is $y = f(x - 2)$

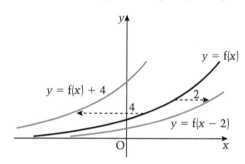

Comparing $y = f(x)$ with $y = f(x - 2)$, we see that the values of y are the same when the value of x in $f(x - 2)$ is 2 units greater than the value of x in $f(x)$.

Therefore for equal values of y, points on $y = f(x - 2)$ are 2 units to the right of points on $y = f(x)$, i.e. the curve $y = f(x - 2)$ is a translation by 2 units of the curve $y = f(x)$ in the positive direction of the x-axis.

Using similar reasoning, the curve $y = f(x + 4)$ is a translation of the curve $y = f(x)$ by 4 units in the direction of the negative x-axis.

> **For any function f, the curve whose equation is $y = f(x + c)$ is a translation of the curve $y = f(x)$ by c units parallel to the x-axis. When $c > 0$, the translation is in the negative direction of the x-axis and when $c < 0$, the translation is in the positive direction of the x-axis.**

Reflections

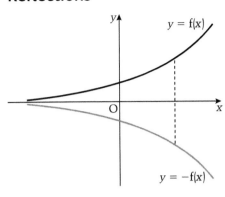

Consider the curve whose equation is $y = -f(x)$

Comparing $y = f(x)$ with $y = -f(x)$, we see that for a given value of x, $f(x) = -f(x)$

Therefore for the same value of x, a point on $y = -f(x)$ is the reflection in the x-axis of the point on $y = f(x)$

So, for any function f, the curve $y = -f(x)$ is the reflection of the curve $y = f(x)$ in the x-axis.

Consider the curve whose equation is $y = f(-x)$

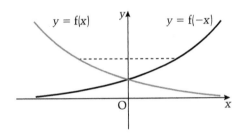

Comparing $y = f(x)$ with $y = f(-x)$, we see that the values of y are equal when the values of x are opposite in sign, i.e. $f(a) = f(-(-a))$

Therefore points with the same y-coordinates on the curves are symmetrical about the y-axis.

So for any function f, the curve $y = f(-x)$ is the reflection of the curve $y = f(x)$ in the y-axis.

Example

Sketch the curve whose equation is $y = \dfrac{1}{x - 2}$

Start with the curve $y = \dfrac{1}{x}$ whose shape and position is known.

If $f(x) = \dfrac{1}{x}$ then $\dfrac{1}{x - 2}$ is $f(x - 2)$.

So the curve $y = \dfrac{1}{x - 2}$ is a translation of $y = \dfrac{1}{x}$ by 2 units in the positive direction of the x-axis.

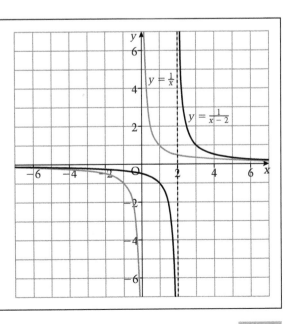

Example

Sketch the curve $y = 2 - (x + 5)^2$

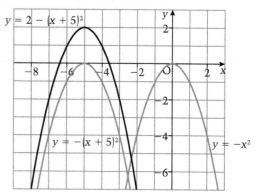

Start with $y = -x^2$ whose shape and position is known.

Then $y = -(x + 5)^2$ is a translation of $y = -x^2$ by 5 units in the direction of the negative x-axis.

Therefore $y = 2 - (x + 5)^2$ is a translation of $y = -(x + 5)^2$ by 2 units parallel to the positive y-axis.

Exercise 1.11a

1   Sketch each of the following curves whose equations are

    (a) $y = x^2 - 4$ **(c)** $y = (x - 1)^3$

    (b) $y = \dfrac{1}{x} + 2$ **(d)** $y = 3 - \dfrac{1}{x - 1}$

2   On the same set of axes, sketch the curves whose equations are

    (a) $y = x^3$ **(c)** $y = -(x + 2)^3$

    (b) $y = (x + 2)^3$ **(d)** $y = 1 - (x + 2)^3$

One-way stretches

Consider the curve whose equation is $y = af(x)$

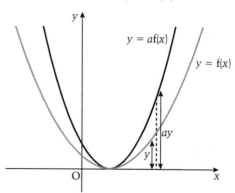

Comparing points on $y = f(x)$ and $y = af(x)$ with the same x-coordinate, the y-coordinate of the point on $y = af(x)$ is a times the y-coordinate on $y = f(x)$

**So the curve $y = af(x)$ is a one-way stretch
of the curve $y = f(x)$ parallel to the y-axis by a factor a.**

Consider the curve whose equation is $y = f(ax)$

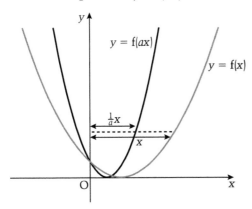

Comparing points on $y = f(x)$ and $y = f(ax)$ with the same y-coordinate, the x-coordinate of the point on $y = f(ax)$ is $\frac{1}{a}$ times the x-coordinate on $y = f(x)$

> **So the curve $y = f(ax)$ is a one-way stretch of the curve $y = f(x)$ parallel to the x-axis by a factor $\frac{1}{a}$.**

Example

On the same set of axes sketch the curves $y = x^2$, $y = 2x^2$ and $y = (2x)^2$ for values of x from -3 to 3.

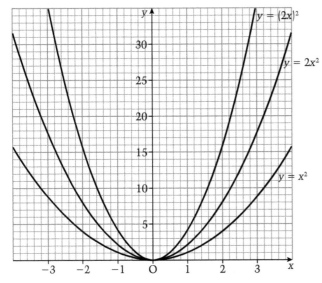

Start with $y = x^2$

Then double the y-coordinate of points on $y = x^2$ to give $y = 2x^2$

Halve the x-coordinate of points on $y = x^2$ to give $y = (2x)^2$

Exercise 1.11b

On the same set of axes sketch the graphs of the curves given for values of x from -3 to 3.

(a) $y = \frac{1}{x}$ **(b)** $y = \frac{2}{x}$ **(c)** $y = \frac{1}{2x}$

Learning outcomes

- To express an improper rational expression as the sum of a polynomial and a proper rational expression

You need to know

- The meaning of the order of a polynomial
- The factor theorem
- How to use transformations to sketch curves

Rational expressions

An expression where both the numerator and denominator are polynomials is called a **rational expression**.

For example, $\dfrac{1}{x}$, $\dfrac{x}{(x-1)(x-2)}$, $\dfrac{3x}{x^2+1}$ are rational expressions.

These expressions are all proper rational expressions because the order of the numerator is less than the order of the denominator.

When the order of the numerator is greater than or equal to the order of the denominator, the expression is called improper.

For example, $\dfrac{x^2}{2x-1}$ and $\dfrac{x+1}{2x-1}$ are **improper**.

Expressing an improper fraction as a sum of a polynomial and a proper fraction

There are two methods we can use to express an improper fraction in a form where the remaining fraction is proper.

The first method involves rearranging the numerator so that we can cancel.

For example, in the case of $\dfrac{2x+3}{x-1}$ we can rearrange the numerator so

that $x - 1$ is part of the numerator,

i.e. $\dfrac{2x+3}{x-1} = \dfrac{2(x-1)+2+3}{x-1}$

We can now express the right-hand side as the sum of two fractions,

i.e. $\dfrac{2(x-1)}{x-1} + \dfrac{5}{x-1}$

We can now cancel $(x-1)$ in the first fraction to give $2 + \dfrac{5}{x-1}$

$\therefore \dfrac{2x+3}{x-1} = 2 + \dfrac{5}{x-1}$

The second method involves dividing the numerator by the denominator.

$$
\begin{array}{r}
2 \\
x-1\overline{\smash{)}\,2x+3} \\
\underline{2x-2} \\
5
\end{array}
$$

Start by dividing x into $2x$. It goes 2 times.
Multiply $x - 1$ by 2 then subtract this from $2x + 3$
2 is the quotient and 5 is the remainder.

Then $\dfrac{2x+3}{x-1} = 2 + \dfrac{5}{x-1}$ $\left(\text{in the same way as } \dfrac{12}{7} = 1 + \dfrac{5}{7}\right)$

This second method is useful when the denominator is quadratic.

For example, to express $\dfrac{x^3 - 2x + 5}{x^2 + 4x + 5}$ in a form where the fraction is

proper, we divide by the denominator.

$$\begin{array}{r} x - 4 \\ x^2 + 4x + 5\overline{)x^3 + 0x^2 - 2x + 5} \\ \underline{x^3 + 4x^2 + 5x} \\ -4x^2 - 7x + 5 \\ \underline{-4x^2 - 16x - 20} \\ 9x + 25 \end{array}$$

There is no x^2 term in the numerator so we add zero for this term. Start by dividing x^2 into x^3. It goes x times. Multiply $x^2 + 4x + 5$ by x then subtract. Bring down 5. Divide x^2 into $-4x^2$, and repeat the process until no more division is possible.

$$\therefore \frac{x^3 - 2x + 5}{x^2 + 4x + 5} = x - 4 + \frac{9x + 25}{x^2 + 4x + 5}$$

This second method of division is also useful when, given one factor of a polynomial, we need to find the other factor and finding it by inspection is not straightforward.

Example

Given that $2x - 1$ is a factor of $2x^4 - x^3 + 6x^2 - x - 1$, find the cubic factor.

$$\begin{array}{r} x^3 + \quad\quad 3x + 1 \\ 2x - 1\overline{)2x^4 - x^3 + 6x^2 - x - 1} \\ \underline{2x^4 - x^3} \\ 0 + 6x^2 - x - 1 \\ \underline{6x^2 - 3x} \\ 2x - 1 \\ \underline{2x - 1} \end{array}$$

Therefore $x^3 + 3x + 1$ is the cubic factor.

Example

Sketch the curve whose equation is $y = \dfrac{x}{x - 1}$

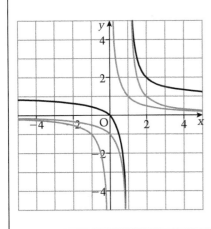

$$\frac{x}{x - 1} = \frac{x - 1 + 1}{x - 1} = 1 + \frac{1}{x - 1}$$

$$\therefore y = 1 + \frac{1}{x - 1}$$

Start with $y = \dfrac{1}{x}$, then $y = \dfrac{1}{x - 1}$ is the translation of $y = \dfrac{1}{x}$ by 1 unit in the direction of the positive x-axis.

So $y = 1 + \dfrac{1}{x - 1}$ is the translation of $y = \dfrac{1}{x - 1}$ by 1 unit parallel to the positive y-axis.

Exercise 1.12

1   Express each expression in a form where the fraction is proper.

   (a) $\dfrac{2x}{2x + 1}$ (b) $\dfrac{6x}{2x - 1}$ (c) $\dfrac{x^3 + x^2 + 3}{x^2 + 4}$

2   Sketch the curve whose equation is $y = \dfrac{2x}{2x + 1}$

3   Show that $x - 2$ is a factor of $x^4 - x^3 - 2x^2 + 3x - 6$
    Hence factorise $x^4 - x^3 - 2x^2 + 3x - 6$

1.13 Inequalities – quadratic and rational expressions

Learning outcomes

- To revise quadratic inequalities
- To solve inequalities involving rational expressions

You need to know

- How to sketch a curve whose equation has the form $y = ax^2 + bx + c$
- The conditions for a quadratic equation to have real roots or no real roots
- How to complete the square

Quadratic inequalities

A quick sketch is the easiest way to solve an inequality such as $(x + 1)(x - 2) < 0$

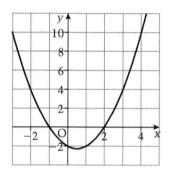

The sketch of the curve $y = (x + 1)(x - 2)$ shows that the curve is below the x-axis, i.e. $(x + 1)(x - 2) < 0$, when $-1 < x < 2$

The inequality can also be solved algebraically.

We know that $(x + 1)(x - 2) = 0$ when $x = -1$ and $x = 2$, and whether $(x + 1)(x - 2)$ is positive or negative depends on the signs of $(x + 1)$ and $(x - 2)$.
So we investigate these signs for $x < -1$, $-1 < x < 2$ and $x > 2$

When $x < -1$, both brackets are negative, so $(x + 1)(x - 2)$ is positive.

When $-1 < x < 2$, $(x + 1)$ is positive and $(x - 2)$ is negative, so $(x + 1)(x - 2)$ is negative.

When $x > 2$, both brackets are positive, so $(x + 1)(x - 2)$ is positive.

Therefore $(x + 1)(x - 2) < 0$ when $-1 < x < 2$

Example

Find the values of a for which $x^2 + ax + a > 0$ for $x \in \mathbb{R}$.

The curve $y = x^2 + ax + a$ is a parabola with a minimum value.

Completing the square gives $x^2 + ax + a = \left(x + \frac{a}{2}\right)^2 + a - \frac{a^2}{4}$

$\left(x + \frac{a}{2}\right)^2 > 0$ for all values of x,

so for $x^2 + ax + a > 0$, $a - \frac{a^2}{4} > 0$

Now, $a - \frac{a^2}{4} > 0 \Rightarrow 4a - a^2 > 0 \Rightarrow a(a - 4) < 0$

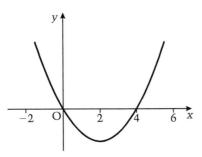

A sketch of $y = a(a - 4)$ shows that $a(a - 4) < 0$ when $0 < a < 4$

Therefore $x^2 + ax + a > 0$ when $0 < a < 4$

Rational expressions

An expression where both the numerator and denominator are polynomials is called a rational expression.

For example,

$\dfrac{1}{x}$, $\dfrac{x}{(x - 1)(x - 2)}$, $\dfrac{3x}{x^2 + 1}$ are rational expressions.

The range of values that a rational function can take

The graph of $y = \dfrac{1}{x}$ (see Topic 1.10), shows that 0 is the only value that y cannot take.

We can show this algebraically: $y = \dfrac{1}{x} \Rightarrow x = \dfrac{1}{y}$ and x is undefined

when $y = 0$, so there is no value of x for which $y = 0$,

i.e. $\dfrac{1}{x} < 0$ or $\dfrac{1}{x} > 0$ but $\dfrac{1}{x} \neq 0$

Now x can take all real values when $y = \dfrac{3x}{x^2 + 1}$

To find the values that y can have, we rearrange the equation to give a quadratic equation in x,

i.e. $yx^2 - 3x + y = 0$

For x to be real, this equation has to have real roots, so '$b^2 - 4ac$' $\geqslant 0$,

i.e. $9 - 4y^2 \geqslant 0$

$\Rightarrow y^2 \leqslant \dfrac{9}{4}$

$\Rightarrow y \geqslant -\dfrac{3}{2}$ and $y \leqslant \dfrac{3}{2}$

$\Rightarrow -\dfrac{3}{2} \leqslant y \leqslant \dfrac{3}{2}$

We can use this information, together with the following observations,

to sketch the graph of $y = \dfrac{3x}{x^2 + 1}$

- $y > 0$ when $x > 0$

- $y = 0$ when $x = 0$

- when $x < 0$, $y < 0$

- as x approaches very large values (we write this as $x \to \infty$), $y \to 0$

- as $x \to -\infty$, $y \to 0$

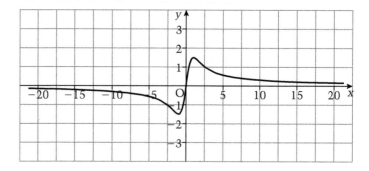

Solving inequalities involving rational expressions

It is easy to see the values of x for which $\dfrac{3x}{x^2 + 1} > 0$, but it is not so obvious for the expression $\dfrac{x}{(x - 1)(x - 2)}$

The values of x for which $\dfrac{x}{(x - 1)(x - 2)} > 0$ depend on the signs of x, $x - 1$ and $x - 2$.

The value of the expression is zero when $x = 0$ and undefined when $x = 1$ and $x = 2$, so we need to investigate the sign of the expression when

$$x < 0, \qquad 0 < x < 1, \qquad 1 < x < 2 \qquad \text{and} \qquad x > 2$$

The easiest way to do this is to use a table.

	$x < 0$	$0 < x < 1$	$1 < x < 2$	$x > 2$
x	−	+	+	+
$x - 1$	−	−	+	+
$x - 2$	−	−	−	+
$\dfrac{x}{(x - 1)(x - 2)}$	−	+	−	+

Now we can see that $\dfrac{x}{(x - 1)(x - 2)} > 0$ when $0 < x < 1$ and $x > 2$

Example

Solve the inequality $\dfrac{x}{2x - 1} < \dfrac{1}{x}$

An inequality is easier to solve if it is first rearranged so that the right-hand side is zero.

$$\frac{x}{2x - 1} < \frac{1}{x}$$

$$\Rightarrow \frac{x}{2x - 1} - \frac{1}{x} < 0$$

$$\Rightarrow \frac{(x - 1)^2}{x(2x - 1)} < 0$$

The numerator is positive for all $x \in \mathbb{R}$ so the significant values of x are 0 and $\frac{1}{2}$.

We need to investigate the ranges $x < 0$, $0 < x < \frac{1}{2}$ and $x > \frac{1}{2}$

	$x < 0$	$0 < x < \frac{1}{2}$	$x > \frac{1}{2}$
$(x - 1)^2$	+	+	+
x	−	+	+
$2x - 1$	−	−	+
$\dfrac{(x - 1)^2}{x(2x - 1)}$	+	−	+

Therefore $\dfrac{x}{2x - 1} < \dfrac{1}{x}$ for $0 < x < \frac{1}{2}$

Example

Find the values of k for which $\dfrac{x^2 - 2x + k}{x + 1}$ can take all real values for all $x \in \mathbb{R}$.

Let $y = \dfrac{x^2 - 2x + k}{x + 1}$

Rearranging as a quadratic equation in x gives
$x^2 - x(2 + y) + (k - y) = 0$

For x to be real, $(2 + y)^2 \geqslant 4(k - y)$

$\Rightarrow \qquad\qquad y^2 + 8y \geqslant (4k - 4)$

Completing the square gives $(y + 4)^2 - 16 \geqslant 4k - 4$

The minimum value of $(y + 4)^2 - 16$ is -16, so y can take all real values provided that $-16 \geqslant 4k - 4$, i.e. $k \leqslant -3$

Exercise 1.13

1   Find the values of x for which $x^2 - 4 < 2x - 1$

2   Find the values of k for which $(kx)^2 + (3k - 2)x + 4 > 0$ for $x \in \mathbb{R}$.

3   Find the set of values of x for which $\dfrac{x - 1}{x(x - 2)} > 0$

4   Find the range of values of y for which $y = \dfrac{2x}{1 + x^2}$

5   Find the minimum value of k for which $\dfrac{x + k}{x^2 + 1} > -1$ for all $x \in \mathbb{R}$.

Learning outcomes

- To investigate the intersection of a curve and a line

You need to know

- How to sketch the graphs of linear, quadratic and cubic functions
- How to solve a pair of simultaneous equations where one is linear and the other is quadratic
- The conditions for a quadratic equation to have two distinct roots, or a repeated root, or no real roots
- How to use the factor theorem to find the roots of a cubic equation

Intersection

Depending on the shape of a curve, a line may intersect the curve at several points, it may touch the curve at one of these points, or it may not intersect the curve at any point.

For example, a line may intersect a parabola in two distinct points, or it may touch the parabola at one point (in which case it is called a tangent to the parabola), or it may not intersect the parabola.

To find the points of intersection, we need to solve the equation of the curve and the equation of the line simultaneously.

For example, to find the points of intersection of the line with equation $y = 3x - 5$ with the curve with equation $y = x^2 - 2x + 1$, we solve the equations simultaneously.

A rough sketch of these curves gives an idea of which of the three cases illustrated above exists in this case. However, this sketch is inconclusive.

The nature of the solution will tell us if this line intersects, touches or misses the curve.

$y = x^2 - 2x + 1$ [1] $y = 3x - 5$ [2]

[2] in [1] $\Rightarrow 3x - 5 = x^2 - 2x + 1 \Rightarrow x^2 - 5x + 6 = 0$

$\Rightarrow (x - 2)(x - 3) = 0$ [3]

There are two real and distinct roots so the line intersects the curve in two distinct points.

From [3], the coordinates of these points are $x = 2$ and (from [2]) $y = 1$, or $x = 3$ and $y = 4$, i.e. (2, 1) and (3, 4).

Example

Prove that the line $y = 3 - 2x$ and the curve $y = \dfrac{2}{x - 3}$ do not intersect.

Solving $y = 3 - 2x$ and $y = \dfrac{2}{x - 3}$ simultaneously gives

$y = 3 - 2x$ [1]

$y = \dfrac{2}{x - 3}$ [2]

Substituting [2] in [1] gives $\dfrac{2}{x - 3} = 3 - 2x$

$\Rightarrow 2 = (3 - 2x)(x - 3)$

$\Rightarrow 2x^2 - 9x + 11 = 0$

'$b^2 - 4ac$' is $81 - 88$ which is less than zero, so there are no real values of x for which $2x^2 - 9x + 11 = 0$

Therefore the line and the curve do not intersect.

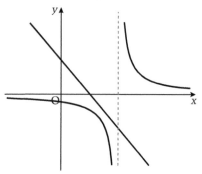

The sketch shows that the line and curve do not intersect but this is not a proof. (A sketch is also unreliable.)

Example

(a) Find the condition on m and c for which the line $y = mx + c$ is a tangent to the curve whose equation is $y = 3x^2 - 2x - 1$

(b) Hence find the equation of the line with gradient 2 that is a tangent to the curve whose equation is $y = 3x^2 - 2x - 1$

(a) Solving $y = 3x^2 - 2x - 1$ and $y = mx + c$ simultaneously gives
$$mx + c = 3x^2 - 2x - 1$$
$$\Rightarrow 3x^2 - x(m + 2) - (c + 1) = 0$$

For the line to touch the curve, this equation must have a repeated root, i.e. '$b^2 - 4ac$' = 0,
so $(m + 2)^2 = 12(c + 1)$

(b) When $m = 2$, $\quad 16 = 12(c + 1) \Rightarrow c = \frac{1}{3}$

$\therefore$ the equation of the line with gradient 2 that is a tangent to

$\quad y = 3x^2 - 2x - 1$ is

$\quad y = 2x + \frac{1}{3}$

$\Rightarrow 3y - 6x - 1 = 0$

Example

Show that the line $y + 5x - 4 = 0$ intersects the curve $y = x^3 - 7x^2 + 10x - 5$ once and touches it once.

The values of x at which the line intersects the curve are given by the roots of the equation

$$x^3 - 7x^2 + 10x - 5 = 4 - 5x \Rightarrow x^3 - 7x^2 + 15x - 9 = 0$$

Possible factors of $f(x) = x^3 - 7x^2 + 15x - 9$ are $(x \pm 1)$, $(x \pm 9)$, $(x \pm 3)$

Using the factor theorem, when $x = 1$, $f(x) = 0$, $\therefore (x - 1)$ is a factor.

So $f(x) = (x - 1)(x^2 - 6x + 9)$
$\quad\quad\quad = (x - 1)(x - 3)^2$

$\therefore$ the equation $x^3 - 7x^2 + 15x - 9 = 0$ has one single root and one repeated root.

Therefore the line $y + 5x - 4 = 0$ intersects the curve $y = x^3 - 7x^2 + 10x - 5$ once and touches it once.

Exercise 1.14

1 Find the value of k for which the line $y = kx + 2$ touches the curve $xy + 4 = 0$

2 Find the nature of the points on the curve $y = x^3 + 5x^2 + 8x + 4$ where $y = 0$. Hence sketch the curve.

3 Determine whether the line $y = x - 5$ intersects, touches or does not intersect the curve whose equation is $x^2 + 2y^2 = 7$

Mappings

When the number 2 is entered in a calculator and then the x^2 button is pressed, the display shows the number 4.
2 is mapped to 4, which is denoted by $2 \mapsto 4$
Under this rule, which is squaring the input number, $3 \mapsto 9$, $25 \mapsto 625$, $0.2 \mapsto 0.04$, $-2 \mapsto 4$ and (any real number) $\mapsto$ (the square of that number).

This is denoted by $x \mapsto x^2$, for $x \in \mathbb{R}$.

This mapping can be represented graphically by plotting values of x^2 against values of x. The graph, and our knowledge of what happens when we square a number, show that one input number gives just one output number.

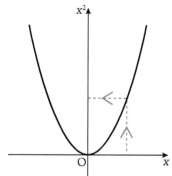

But the mapping that maps a number to its square root, e.g. $4 \mapsto \pm 2$, gives a real output only when the input number is greater than or equal to zero (negative numbers do not have real square roots).

This mapping can be written as $x \mapsto \pm\sqrt{x}$, for $x \in \mathbb{R}$.

The graphical representation of this mapping shows that one input value gives two output values.

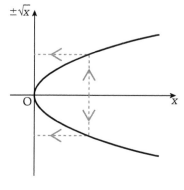

Functions

For the mapping $x \mapsto x^2$, for $x \in \mathbb{R}$, one input number gives one output number.

The mapping $x \mapsto \pm\sqrt{x}$ gives two outputs for every one input number.

The word function is used for any mapping where one input value gives one output value.

> **A *function* is a rule that maps each single number to another single number for a defined set of input numbers.**

Using f for function and the symbol : to mean 'such that', we write $f : x \mapsto x^2$ for $x \in \mathbb{R}$ to mean f is the function that maps x to x^2 for all real values of x.

The mapping $x \mapsto \pm\sqrt{x}$ for $x \geqslant 0$, $x \in \mathbb{R}$ is not a function because it does not satisfy this condition.

Domain and range

We have assumed that we can use any real number as an input for a function unless some particular numbers have to be excluded because they do not give real numbers as output.

The set of inputs for a function is called the ***domain*** of the function.

The domain is also called the pre-image.

The domain does not have to contain all possible inputs; it can be as wide, or as restricted, as we choose to make it. Hence to define a function fully, the domain must be stated.

If the domain is not stated, we assume that it is the set of all real numbers ($\mathbb{R}$).

The mapping $x \mapsto x^2 + 3$ can be used to define a function f over any domain we choose. Some examples, together with their graphs, are given.

1 $f : x \mapsto x^2 + 3$ for $x \in \mathbb{R}$

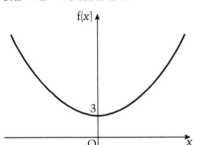

2 $f : x \mapsto x^2 + 3$ for $x \in \mathbb{R}\ x \geqslant 0$

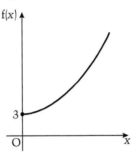

The point on the curve where $x = 0$ is included, and we denote this by a solid circle. For the domain $x > 0$, the point $x = 0$ would not be part of the curve, and we indicate this by using an open circle.

3 $f : x \mapsto x^2 + 3$ for $x = 1, 2, 3, 4$

This time the graphical representation is four discrete points.

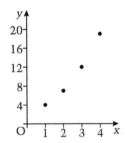

These three examples are not the same function – each is a different function.

For each domain, there is a corresponding set of output numbers.

The set of output numbers is called the ***range*** of the function. The range is also called the image.

The notation $f(x)$ represents the output values of a function, so for $f : x \mapsto x^2$ for x, $f(x) = x^2$

For the function defined in **1** above, the range is $f(x) \geqslant 3$, for the function given in **2**, the range is also $f(x) \geqslant 3$ and for the function defined in **3**, the range is the set of numbers 4, 7, 12, 19.

A function can be represented pictorially.
For example, $f : x \mapsto x^2 + 3$ for $x = 1, 2, 3, 4$ can be illustrated as:

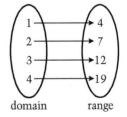

This function can also be represented by a set of ordered pairs, where the first number in the pair is the value of x, and the second number is the value of $f(x)$. Therefore $f : x \mapsto x^2 + 3$ for $x = 1, 2, 3, 4$ can be represented by the set $\{(1, 4), (2, 7), (3, 12), (4, 19)\}$.

Example

The diagram shows a mapping of members of the set A{1, 2, 3, 4, 5} to the set B{a, b, c, d, e}.

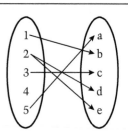

(a) Give two reasons why this mapping is not a function.

(b) Construct a function, f, that maps A to B, giving your answer as a set of ordered pairs.

(a) In A, 2 maps to two different members of B (d and e). In A, 4 does not map to any member of B.

(b) For f to be a function, every member of A must map to just one member of B. There are several ways of doing this but the simplest is to change the two mappings in A so that 2 maps to either d (or e), and then 4 maps to e (or d). f = {(1, b), (2, d), (3, c), (4, e), (5, a)}

Example

The function, f, is defined by $f(x) = x^2$ for $x \leqslant 0$,
and $f(x) = x$ for $x > 0$, $x \in \mathbb{R}$.

(a) Find $f(4)$ and $f(-4)$ **(b)** Sketch the graph of f. **(c)** Give the range of f.

(a) For $x > 0$, $f(x) = x$ $\therefore f(4) = 4$
For $x \leqslant 0$, $f(x) = x^2$ $\therefore f(-4) = (-4)^2 = 16$

(b) To sketch the graph of a function, we can use what we know about lines and curves in the xy-plane. So we can interpret $f(x) = x$ for $x > 0$, as that part of the line $y = x$ which corresponds to positive values of x, and $f(x) = x^2$ for $x \leqslant 0$ as the part of the curve $y = x^2$ that corresponds to negative values of x.

(c) The range of f is $f(x) \geqslant 0$

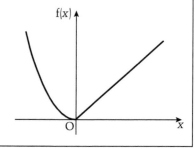

Composite functions

Two functions f and g are given by $f(x) = x^2$, $x \in \mathbb{R}$ and g(x) and $g(x) = \frac{1}{x}$, $x \neq 0$, $x \in \mathbb{R}$. These two functions can be combined in several ways.

1 They can be added or subtracted,

i.e. $f(x) + g(x) = x^2 + \frac{1}{x}$, $x \neq 0$, $x \in \mathbb{R}$

and $f(x) - g(x) = x^2 - \frac{1}{x}$, $x \neq 0$, $x \in \mathbb{R}$

2 They can be multiplied or divided,

i.e. $f(x)g(x) = x^2 \times \frac{1}{x} = x$, $x \neq 0$, $x \in \mathbb{R}$

and $\frac{f(x)}{g(x)} = \frac{x^2}{\frac{1}{x}} = x^3$, $x \neq 0$, $x \in \mathbb{R}$

3 The output of f can be made the input of g,

i.e. $x \overset{f}{\mapsto} x^2 \overset{g}{\mapsto} \frac{1}{x^2}$ or $g[f(x)] = g(x^2)$, $x \neq 0$, $x \in \mathbb{R}$

Therefore the function $f : x \mapsto \frac{1}{x^2}$, $x \neq 0$, $x \in \mathbb{R}$ is obtained by taking the function g of the function f. This is a **composite function** and is written as gf(x).

For $f(x) = x^2$, $x \in \mathbb{R}$ and $g(x) = 3x - 1$, $x \in \mathbb{R}$

gf(x) means the function g of the function f(x),
i.e. $gf(x) = g(x^2) = 3x^2 - 1$, $x \in \mathbb{R}$

fg(x) means the function f of the function g(x),
i.e. fg(x) = f($3x − 1$) = $(3x − 1)^2$, $x \in \mathbb{R}$.

This shows that the composite function fg(x) is not the same as the composite function gf(x).

For any composite function gf(x), f(x) is the range of f and this range gives the input values of g. Therefore the range of f must be included in the domain of g.

Example

f, g and h are functions given by
f(x) = x^2, $x \in \mathbb{R}$, g(x) = $2x + 1$, $x \in \mathbb{R}$, h(x) = $1 − x$, $x \in \mathbb{R}$.

(a) Find as a function of x: **(i)** fg **(ii)** ghf.

(b) Calculate the value of: **(i)** gf(3) **(ii)** hfg(3) **(iii)** gg(3).

(a) **(i)** fg(x) = f($2x + 1$) = $(2x + 1)^2$, $x \in \mathbb{R}$

  (ii) ghf(x) = gh(x^2) = g($1 − x^2$)
  $= 2(1 − x^2) + 1$
  $= 3 − 2x^2$, $x \in \mathbb{R}$

(b) **(i)** gf(x) = g(x^2) = $2x^2 + 1$, ∴ gf(3) = $2(3)^2 + 1 = 19$

  (ii) hfg(x) = hf($2x + 1$) = h($(2x + 1)^2$) = $1 − (2x + 1)^2$,
  ∴ hfg(3) = $1 − (7)^2 = −48$

  (iii) gg(x) = g($2x + 1$) = $2(2x + 1) + 1 = 4x + 3$
  ∴ gg(3) = $4(3) + 3 = 15$

Example

f and g are functions of x such that f(x) = $\dfrac{1}{2x − 1}$ and gf(x) = x
Find g(x).

To change $\dfrac{1}{2x − 1}$ to x, we need to first take the reciprocal of $\dfrac{1}{2x − 1}$,

so let h(x) = $\dfrac{1}{x}$, then hf(x) = $2x − 1$

To change $2x − 1$ to x, we need to halve $2x − 1$ and then add $\frac{1}{2}$,

so let j(x) = $\frac{1}{2}x + \frac{1}{2}$, then jhf($x$) = $\frac{1}{2}(2x − 1) + \frac{1}{2} = x$

jh(x) = $\dfrac{1}{2x} + \dfrac{1}{2}$ = g(x)

Exercise 1.15

1 The function f is given by f(x) = $−x$ for $x < 0$, $x \in \mathbb{R}$
  and f(x) = x for $x \geqslant 0$, $x \in \mathbb{R}$.
  (a) Find the value of f(5), f($−3$) and f(0).
  (b) Sketch the graph of the function.

2 The functions f and g are given by f(x) = x^2, $x \in \mathbb{R}$ and g(x) = $2 − x$
  (a) (i) Find the function given by fg(x).
    (ii) Sketch the curve whose equation is y = fg(x)
  (b) (i) Find the function given by gf(x).
    (ii) Sketch the curve whose equation is y = gf(x)

3 f and g are functions such that f(x) = $\dfrac{3}{2x + 1}$ and gf(x) = x. Find g(x).

Codomain

The **codomain** of a function is the possible values that can come out of a function.

This will include the actual values that come out (i.e. the range) but may include other values as well. The codomain is useful when we do not know all the values that will come out of a function (e.g. a complicated function or one we have not seen before). We can then choose as a codomain the values that might come out of a function.

For the function $f : x \mapsto x^2$ for $x = 1, 2, 3$, we can choose the codomain to be the set of integers from 1 to 10 inclusive.

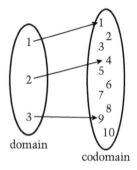

One-to-one functions

A function is **one-to-one** when each member of the domain maps to a different member of the codomain. The function f defined above is a one-to-one function.

A one-to-one function is also called an **injective function**.

Onto functions

A function is **onto** when every member of the codomain is mapped to by a member of the domain, i.e. no members of the codomain are left without a matching member of the domain.

For example, when g is given as $g(x) = x^2$ for $x = -2, -1, 0, 1, 2$ and the codomain is the set $\{0, 1, 4\}$, the diagram shows that every member of the codomain has at least one image in the domain.
Therefore g is an onto function.

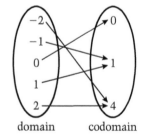

An onto function is also called a **surjective** function.

The function $f : x \mapsto x^2$ for $x = 1, 2, 3$ given above is not surjective because some members of the codomain do not have an image in the domain.

However, $h : x \mapsto x^2$ for $x \in \mathbb{R}$ onto the positive real numbers is surjective because every real number, when it is squared, maps to a positive real number, and every positive real number has a real square root. But h is not one-to-one because more than one member of the domain maps to one member of the codomain, e.g. both 2 and -2 map to 4.

Bijective functions

A function that is both one-to-one and onto is called a *bijective* function. This is a function where each member of the domain maps to one member of the codomain and where each member of the codomain comes from one member of the domain.

For example, $f : x \mapsto 2x$ for $x \in \mathbb{R}$ onto $\mathbb{R}$ is one-to-one (each member of $\mathbb{R}$ maps to one member of $\mathbb{R}$) and onto (each member of $\mathbb{R}$ comes from just one member of $\mathbb{R}$). Therefore f is a bijective function.

To summarise:

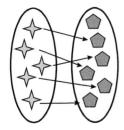   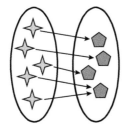   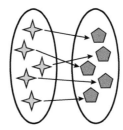

injective but not surjective	**surjective but not injective**	**bijective**
Every member of the domain maps to a different member of the codomain.	Some members of the domain map to the same member of the codomain and every member of codomain is mapped to.	Every member of the domain maps to a different member of the codomain and every member of the codomain is mapped to.

Example

Determine whether the function given by $f(x) = x^2 - 2$, $x \in \mathbb{R}$ onto the codomain $\mathbb{R}$ is injective, surjective or neither.

$x^2 - 2 \geqslant -2$ for all $x \in \mathbb{R}$, therefore some members of the codomain are not mapped to. Therefore f is not surjective.

When $x = -3$ and $x = 3$, $f(x) = 7$, therefore more than one member of the domain maps to one member of the codomain.

Therefore f is not injective.

So $f(x) = x^2 - 2$, $x \in \mathbb{R}$ onto the codomain $\mathbb{R}$ is neither injective nor surjective.

Exercise 1.16

1 Let $A = \{1, 2, 3, 4\}$ and the function $f : A \mapsto A$ be given by
$f = \{(1, 2), (2, 4), (3, 1), (4, 3)\}$

Show that f is one-to-one and onto.

2 Let $A = \{-1, 0, 1, 2\}$ and the function $f : A \mapsto A$ be given by
$f = \{(-1, 1), (0, 0), (1, 1), (2, 4)\}$

(a) Show that f is not one-to-one.

(b) Is f an onto function? Give a reason for your answer.

Learning outcomes

- To define mathematically the term inverse function

You need to know

- The meaning of a function
- What a one-to-one function is
- The shape of a curve when its equation is $y =$ a quadratic function of x or a cubic function of x

Inverse functions

f is the function where $f(x) = 2x$ for $x = 2, 3, 4$

The domain $\{2, 3, 4\}$ maps to the range $\{4, 6, 8\}$.

The mapping can be reversed, i.e. each member of the range can be mapped back to the domain.

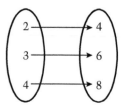

We can express this reverse mapping as
$x \mapsto \frac{1}{2}x$ for $x = 4, 6, 8$

This is a function in its own right.

It is called the ***inverse function*** of f where $f(x) = 2x$

Denoting the inverse function of f by f^{-1}, we write $f^{-1}(x) = \frac{1}{2}x$ for $x = 4, 6, 8$

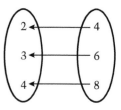

Notice that the range of the function is the domain of the inverse function.

Not every function has an inverse.

Consider the function $f(x) = x^2$ for $x \in \mathbb{R}$, which is such that $+2$ and -2, for example, both map to 4. When this mapping is reversed, each value of x^2 maps to two values of x, for example 4 maps to both $+2$ and -2, and this is not a function. Hence only functions where each member of the domain maps to a different member of the codomain have an inverse, i.e. only one-to-one functions have an inverse.

A function f has an inverse only if f is a one-to-one and an onto function.

You can tell whether a function $f(x)$ is one-to-one from the graph of $y = f(x)$

When any line parallel to the x-axis will cut the graph only once, f is one-to-one.

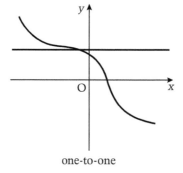 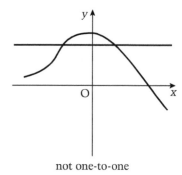

one-to-one   not one-to-one

The graph of a function and its inverse

The diagram shows the curve that is obtained by reflecting $y = f(x)$ in the line $y = x$

The reflection of a point A(a, b) on the curve $y = f(x)$ is the point A' whose coordinates are (b, a), i.e. interchanging the x- and y-coordinates of A gives the coordinates of A'.

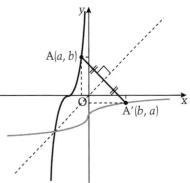

Therefore we can obtain the equation of the reflected curve by interchanging x and y in the equation $y = f(x)$

Now the coordinates of A on $y = f(x)$ can be written as $[a, f(a)]$. Therefore the coordinates of A' on the reflected curve are $[f(a), a]$, i.e. the equation of the reflected curve is such that the output of f is mapped to the input of f.

Hence if the equation of the reflected curve can be written in the form $y = g(x)$, then g is the inverse of f, i.e. $g = f^{-1}$

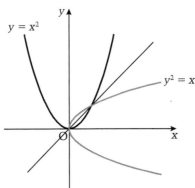

Any curve whose equation can be written in the form $y = f(x)$ can be reflected in the line $y = x$

However, this reflected curve may not have an equation that can be written in the form $y = f^{-1}(x)$

The diagram shows the curve $y = x^2$ and its reflection in the line $y = x$

The equation of the image curve is $x = y^2 \Rightarrow y = \pm\sqrt{x}$ and $x \mapsto \pm\sqrt{x}$ is not a function.

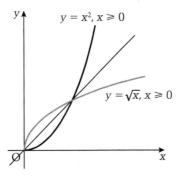

(We can see this from the diagram as, on the reflected curve, one value of x maps to two values of y. So in this case y cannot be written as a function of x.)

However, if we change the domain to give the function $f(x) = x^2$, $x \geqslant 0$, $x \in \mathbb{R}$, then f is a one-to-one function so does have an inverse.

Example
Find $f^{-1}(4)$ when $f(x) = 5x - 1$, $x \in \mathbb{R}$.

$y = 5x - 1$

For the reflected curve $x = 5y - 1 \Rightarrow y = \frac{1}{5}(x + 1)$ and $\frac{1}{5}(x + 1)$ is a function, so $f^{-1}(x) = \frac{1}{5}(x + 1)$, $\therefore f^{-1}(4) = \frac{1}{5}(4 + 1) = 1$

Exercise 1.17

1 A function f is defined by $f : x \mapsto (3 - x)^2$, $x < 3$, $x \in \mathbb{R}$. Define $f^{-1}(x)$ fully.

2 The functions f and g are given by $f(x) = 2x$, $x \in \mathbb{R}$ and $g(x) = 2 - x$, $x \in \mathbb{R}$.

Find: **(a)** $g^{-1}f^{-1}(x)$ **(b)** $(gf)^{-1}(x)$

3 **(a)** Show that $f(x) = (x - 1)(x - 2)(x - 3)$, $x \in \mathbb{R}$ does not have an inverse.

(b) Redefine $f(x)$ with a different domain, so that $f^{-1}(x)$ exists.

1.18 Logarithms

Learning outcomes

- To use the laws of logarithms to simplify expressions

You need to know

- The value of simple powers of 2, 3, 5

Indices

Logarithms depend on the laws of indices, so here is a reminder of these laws.

- $a^p \times a^q = a^{p+q}$ For example, $x^3 \times x^4 = x^{3+4} = x^7$
- $a^p \div a^q = a^{p-q}$ For example, $x^3 \div x^4 = x^{3-4} = x^{-1}$
- $(a^p)^q = a^{pq}$ For example, $(x^3)^4 = x^{3 \times 4} = x^{12}$
- $a^0 = 1,\ a^{-n} = \dfrac{1}{a^n},\ a^{\frac{1}{n}} = \sqrt[n]{a}$

Logarithms

We can read the statement $10^2 = 100$ as

the base 10 raised to the power 2 gives 100.

This relationship can be rearranged to give the same information, i.e.

2 is the power to which the base 10 must be raised to give 100.

In this form the power is called a *logarithm*.

The whole relationship can then be abbreviated to read

2 is the logarithm to the base 10 of 100

or $2 = \log_{10} 100$

In the same way, $2^3 = 8 \Rightarrow 3 = \log_2 8$

and $3^4 = 81 \Rightarrow 4 = \log_3 81$

Similarly, $\log_5 25 = 2 \Rightarrow 25 = 5^2$

and $\log_9 3 = \frac{1}{2} \Rightarrow 3 = 9^{\frac{1}{2}}$

The base of a logarithm can be any positive number, so

$$b = a^c \Leftrightarrow \log_a b = c,\ a > 0$$

The symbol $\Leftrightarrow$ means that each of these facts implies the other.

Also, as $a^0 = 1$ this means $\log_a 1 = 0$,
i.e. **the logarithm of 1 to any base is zero.**

The power of a positive number always gives a positive result,
e.g. $4^2 = 16,\ 4^{-2} = \frac{1}{16},\ \ldots$

This means that, if $\log_a b = c$, i.e. $b = a^c$, then b must be positive. So logs of positive numbers exist,
but **the logarithm of a negative number does not exist.**

Natural logarithms

There is an irrational number that appears in several different areas of mathematics. It is denoted by e and is equal to 2.71828...

This constant was first named e by Euler who showed that as

$x \to \infty, \left(x + \dfrac{1}{x}\right)^x \to e$

Newton discovered that the sum $1 + \dfrac{1}{1} + \dfrac{1}{1 \times 2} + \dfrac{1}{1 \times 2 \times 3}$

$+ \dfrac{1}{1 \times 2 \times 3 \times 4} + \ldots \to e$ as more and more terms are added.

When e is used as the base for logarithms they are called natural logarithms and are denoted by $\ln x$.

$$\ln x \text{ means } \log_e x \quad \text{so} \quad \ln x = y \Leftrightarrow e^y = x$$

Logarithms with a base of 10 are called common logarithms and denoted by $\lg x$ or $\log x$, i.e. if the base is not given, it is taken to be 10. So $\log x$ means $\log_{10} x$ and $\log x = y \Leftrightarrow 10^y = x$

Evaluating logarithms

A scientific calculator can be used to find the values of logarithms with base e or 10.

Use the 'ln' button to evaluate natural logarithms and the 'log' button to evaluate common logarithms. The e^x button (usually above the 'ln' button) is used to evaluate powers of e.

Laws of logarithms

Given $x = \log_a b$ and $y = \log_a c$ then $a^x = b$ and $a^y = c$

Now $bc = (a^x)(a^y) \Rightarrow bc = a^{x+y}$

Therefore $\log_a bc = x + y$ i.e. $\log_a bc = \log_a b + \log_a c$

This is the first law of logarithms and, as a can represent *any* base, this law applies to the logarithm of *any* product *provided that the same base is used for all the logarithms in the formula*.

Using x and y again, a law for the log of a fraction can be found.

$$\frac{b}{c} = \frac{a^x}{a^y} \Rightarrow \frac{b}{c} = a^{x-y}$$

Therefore $\log_a\left(\frac{b}{c}\right) = x - y$ i.e. $\log_a\left(\frac{b}{c}\right) = \log_a b - \log_a c$

A third law allows us to deal with an expression of the type $\log_a b^n$.

Using $x = \log_a b^n \Rightarrow a^x = b^n$ i.e. $a^{\frac{x}{n}} = b$

Therefore $\frac{x}{n} = \log_a b \Rightarrow x = n\log_a b$ i.e. $\log_a b^n = n\log_a b$

These are the most important laws of logarithms. Because they are true for *any* base we do not include a base, but in each of these laws every logarithm must be to the same base.

$$\log bc = \log b + \log c, \quad \log\frac{b}{c} = \log b - \log c, \quad \log b^n = n\log b$$

Example

Express $\log pq^2\sqrt{r}$ in terms of the simplest possible logarithms.

$\log pq^2\sqrt{r}$
$= \log p + \log q^2 + \log\sqrt{r}$
$= \log p + 2\log q + \frac{1}{2}\log r$

Example

Express
$3\log p + n\log q - 4\log r$
as a single logarithm.

$3\log p + n\log q - 4\log r$
$= \log p^3 + \log q^n - \log r^4$
$= \log\dfrac{p^3 q^n}{r^4}$

Exercise 1.18

1 Find the value of:
   (a) $\log_2 16$ **(b)** $\log_4 2$ **(c)** $\log_4 8$

2 Express in terms of the simplest possible logarithms:
   (a) $\log\dfrac{p}{q}$ **(b)** $\ln 5x^2$ **(c)** $\log p\sqrt{q}$ **(d)** $\ln\dfrac{x}{x+1}$

3 Express as a single logarithm:
   (a) $\log p - \log q$ **(c)** $2\log p + 5\log q$
   (b) $\ln 3 + \ln x$ **(d)** $2\ln x - \frac{1}{4}\ln(x-1)$

1.19 Exponential and logarithmic equations

Learning outcomes

- To solve logarithmic and exponential equations including changing the base of a logarithm

You need to know

- The laws of logarithms
- How to simplify logarithms

Exponential equations

An exponential equation has x as part of the index, for example, $3^{x-2} = 8$

When you need to solve an exponential equation, first look to see if the solution is obvious.

For example, for $5^{2-x} = 125$, notice that $125 = 5^3$

Therefore $\qquad 5^{2-x} = 5^3 \quad$ so $2 - x = 3 \Rightarrow x = -1$

When the solution is not obvious, taking logarithms of both sides can change the index to a factor.

For example, for $3^{x-3} = 8$, taking logs of both sides gives

$$(x - 3)\log 3 = \log 8 \quad \text{so} \quad x - 3 = \frac{\log 8}{\log 3}$$

Therefore $x - 3 = 1.892... \Rightarrow x = 4.89$ (3 s.f.)

Note that $\dfrac{\log 8}{\log 3}$ is NOT equal to $\log \dfrac{8}{3}$. Note also that we could equally well have used natural logarithms.

Example

Solve the equation $2^x + 2(2^{-x}) = 3$

The left-hand side of this equation cannot be simplified so taking logs will not help, but using $y = 2^x$ will.

Let $y = 2^x$, then $2^x + 2(2^{-x}) = 3 \Rightarrow y + \dfrac{2}{y} = 3 \Rightarrow y^2 - 3y + 2 = 0$

$\therefore (y - 2)(y - 1) = 0 \Rightarrow y = 1$ or $y = 2$

So $2^x = 1 \Rightarrow x = 0$ or $2^x = 2 \Rightarrow x = 1$

Example

Solve the equation $4(3^x)(5^x) = 7$

$4(3^x)(5^x) = 7 \Rightarrow (3^x)(5^x) = 1.75$

Taking logs gives $\ln (3^x)(5^x) = \ln 1.75 \Rightarrow x \ln 3 + x \ln 5 = \ln 1.75$

$\therefore x = \dfrac{\ln 1.75}{\ln 3 + \ln 5} = 0.207$ (3 s.f.)

Logarithmic equations

A logarithmic equation contains the logarithms of expressions containing x, for example, $\ln (x - 2) = 1 - \ln x$

To solve a logarithmic equation, again look to see if the solution is obvious.

For example, for $\log_2 (2x - 1) = 2 \log_2 x$, we can write $2 \log_2 x$ as $\log_2 x^2$,
then $\qquad\qquad \log_2 (2x - 1) = \log_2 x^2 \Rightarrow 2x - 1 = x^2$

$\qquad\qquad\qquad\qquad\qquad \Rightarrow x^2 - 2x + 1 = 0 \Rightarrow x = 1$

When the solution is not obvious, express the logarithmic terms as a single logarithm and then remove the logarithm.

For example, for $3 \log_2 x = \log_2 16 - 1$, collecting the logarithmic terms on one side and expressing as a single term gives $\log_2 \dfrac{x^3}{16} = -1$, then removing the log gives $\dfrac{x^3}{16} = \dfrac{1}{2}$

Therefore $x^3 = 8 \Rightarrow x = 2$

Example

Solve the equation $\ln x - 2 = \ln(x - 1)$

$\ln x - 2 = \ln(x + 1) \Rightarrow \ln x - \ln(x + 1) = 2$

$$\Rightarrow \ln \dfrac{x}{x + 1} = 2$$

$$\Rightarrow \dfrac{x}{x + 1} = e^2$$

So $x(1 - e^2) = e^2 \Rightarrow x = \dfrac{e^2}{1 - e^2} = -1.16$ (3 s.f.)

Changing the base of a logarithm

When the bases of logarithmic terms are different, they cannot be simplified to a single logarithmic term. To do that, we need to be able to change the base of the logarithm.

If $x = \log_a c$ and we want to change the base of the logarithm to b, then
$x = \log_a c \Rightarrow c = a^x$

Taking logarithms to the base b gives $\log_b c = x \log_b a \Rightarrow x = \dfrac{\log_b c}{\log_b a}$

$$\text{i.e. } \log_a c = \dfrac{\log_b c}{\log_b a} \quad \text{and in particular} \quad \log_a c = \dfrac{\ln c}{\ln a}$$

The base of an exponential expression can change in a similar way.

To express a^x as a power of e, then using $a^x = e^p$ gives $x \ln a = p$, therefore

$$a^x = e^{x \ln a}$$

Example

Solve the equation $3 \log_3 x = 1 + 2 \log_9 x$

First change the base of $\log_9 x$ to 3.

$\log_9 x = \dfrac{\log_3 x}{\log_3 9} = \dfrac{\log_3 x}{2}$,

$\therefore \ 3 \log_3 x = 1 + 2 \log_9 x \Rightarrow 3 \log_3 x = 1 + \log_3 x$

$2 \log_3 x = 1$ so $\log_3 x = \dfrac{1}{2} \Rightarrow x = 3^{\frac{1}{2}} = 1.73$ (3 s.f.)

Exercise 1.19

1　Solve the equation $(4^{2x + 1})(5^{x - 2}) = 6^{1 - x}$

2　Solve the simultaneous equations
$\ln(x + 1) + \ln 2 = \ln y$ and $\ln(x - 2y + 1) = 0$

3　Solve the equation $\log_2 x + \log_x 2 = 2$

4　Given that $\ln y = 3$, find the value of x given that $\ln x^3 + 4 \log_y 5 = 8$

Learning outcomes

- To define exponential and logarithmic functions

You need to know

- The definition of a function and the meaning of a one-to-one function
- How to find an inverse function
- The meaning of logarithms and the laws of logarithms
- The meaning of natural logarithms

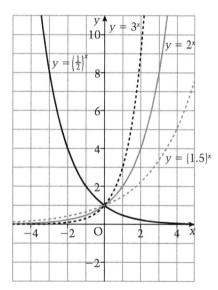

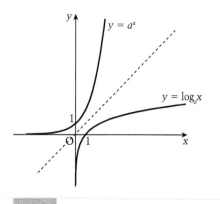

Exponential functions

The mapping $x \mapsto 2^x$ is such that $2 \mapsto 2^2 = 4$, $-2 \mapsto 2^{-2} = \frac{1}{4}$, and any real number maps to a single real number.
Therefore $x \mapsto 2^x$, $x \in \mathbb{R}$ is a function.

But $x \mapsto (-2)^x$ has a real value only when x is an integer, so $(-2)^x$, $x \in \mathbb{R}$ is not a function.

However, for any value of $a > 0$, $x \mapsto a^x$, $x \in \mathbb{R}$ is a function.

The function $f(x) = a^x$, $x \in \mathbf{R}$ is called an exponential function.

For all values of $a > 0$, $a^x > 0$, therefore the range of $f(x) = a^x$, $x \in \mathbb{R}$ is $f(x) > 0$

The curve $y = a^x$

The family of curves whose equations are $y = a^x$ go through a point that is common to all of them: when $x = 0$, $a^0 = 1$, i.e. they all go through the point $(0, 1)$.

When $a = 1$, $y = 1^x = 1$

When $a > 1$, and $x > 0$, a^x increases as x increases
$\qquad$ (e.g. 2^2, 2^3, ..., 2^{10}, ...)

$\qquad$ and when $x < 0$, a^x decreases as x decreases
$\qquad$ (e.g. 2^{-2}, 2^{-3}, ..., 2^{-10}, ...), but never reaches 0, i.e. as $x \to -\infty$, $a^x \to 0$

When $a < 1$, the opposite happens: as $x \to -\infty$, a^x increases
$\qquad$ (e.g. $\left(\frac{1}{2}\right)^{-2} = 4$, ... $\left(\frac{1}{2}\right)^{-5} = 32...$)
$\qquad$ as $x \to \infty$, $a^x \to 0$

This graph shows the curve $y = a^x$ for some different values of x.

The inverse of the function $f(x) = a^x$

The function $f(x) = a^x$ is a one-to-one function, so it has an inverse.

If $y = a^x$ where $f(x) = a^x$, we obtain the graph of $y = f^{-1}(x)$ by reflecting $y = a^x$ in the line $y = x$

We can obtain the equation of this reflected curve by interchanging x and y, so the equation of $y = f^{-1}(x)$ is given by $x = a^y$. Taking logarithms to base a, we get $\log_a x = y$, i.e. $y = \log_a x$

Therefore when $f(x) = a^x$, $f^{-1}(x) = \log_a x$

The function $\log_a x$ has domain $x > 0$.
(The range of a function is the domain of the inverse function.)

The function $f: x \mapsto \log_a x$, $x > 0$, $x \in \mathbf{R}$ is called a logarithmic function and it is the inverse of $f: x \mapsto a^x$, $x \in \mathbf{R}$

The graph shows the curves with equations $y = a^x$ and $y = \log_a x$

The functions e^x and $\ln x$

When $a = e$,

> $f(x) = e^x, x \in \mathbf{R}$ is called the exponential function and
> $f(x) = \ln x, x > 0, x \in \mathbf{R}$ is called the logarithmic function.
>
> **The logarithmic function is the inverse of the exponential
> function and vice versa.**

The graph shows the curves $y = e^x$ and $y = \ln x$

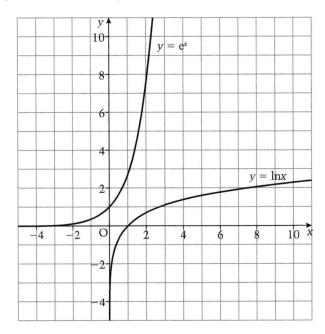

Note that the x-axis is an asymptote to the curve $y = e^x$
and the y-axis is an asymptote to the curve $y = \ln x$

These sketches show some simple variations of the graph of $y = e^x$

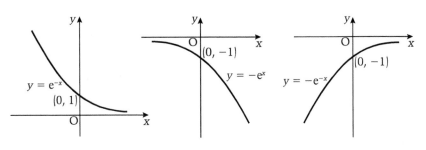

Example

Given $f(x) = e^{3x} + 1, x \in \mathbf{R}$,
find $f^{-1}(x)$.

When $y = e^{3x} + 1$,
interchanging x and y gives
$$x = e^{3y} + 1$$
$$\Rightarrow e^{3y} = x - 1$$
$$\Rightarrow 3y = \ln(x - 1)$$

Therefore $f^{-1}(x) = \frac{1}{3}\ln(x - 1)$,
$x > 1, x \in \mathbf{R}$.

Exercise 1.20

1 On the same set of axes, sketch the graphs of
$y = 1 + e^x, y = 1 - e^x, y = 1 - e^{-x}$

2 On the same set of axes, sketch the graphs of
$y = 1 - \ln x, y = \ln x - 1, y = \ln(x - 1)$

3 Given $f(x) = 1 - e^{-x}, x \in \mathbf{R}$, find $f^{-1}(4)$.

4 Given $f(x) = 1 - 2\ln x$, find $f^{-1}(2)$.

- To define the modulus function and its properties

- How to sketch simple curves
- Algebraic methods for solving inequalities

The modulus of x

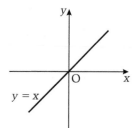

The modulus of x is written as $|x|$ and it means the positive value of x whether or not x itself is positive or negative, e.g. $|2| = 2$ and $|-2| = 2$

Hence the graph of $y = |x|$ can be found from the graph of $y = x$ by changing the part of the graph for which y is negative to the equivalent positive values, i.e. by reflecting the part of the graph where y is negative in the x-axis.

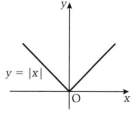

Hence we define the function $f : x \mapsto |x|$,

as $\left.\begin{array}{l} |x| = x \text{ for } \geqslant 0 \\ |x| = -x \text{ for } x < 0 \end{array}\right\} x \in \mathbb{R}$

The properties of $|x|$

$|x|$ is always positive or zero, so we can write $|x| =$ the positive square root of x^2, i.e. $|x| = \sqrt{x^2}$

Now for any two real numbers x and y,

$|x| = x$ for $x \geqslant 0$ and $|x| = -x$ for $x < 0$, and $|y| = y$ for $y \geqslant 0$ and $|y| = -y$ for $y < 0$

If $|x| = |y|$ then $x = \pm y$ so $x^2 = y^2$ and conversely if $x^2 = y^2$, then $x = \pm y$ so $|x| = |y|$,

i.e. $\qquad |x| = |y| \Leftrightarrow x^2 = y^2$

It follows that $|x| < |y| \Leftrightarrow x^2 < y^2$

Now $x^2 < y^2 \Rightarrow x^2 - y^2 < 0$
$\Rightarrow (x - y)(x + y) < 0$:

	$x < -y$	$-y < x < y$	$x > y$
$(x - y)$	$-$	$-$	$+$
$(x + y)$	$-$	$+$	$+$

The table shows that $x^2 < y^2$ when $-y < x < y$,

i.e. $\qquad |x| < |y| \Leftrightarrow -y < x < y$

The last property is $|x + y| \leqslant |x| + |y|$

We can illustrate this on a diagram.

The modulus of a number is equal to the distance of the point representing that number from zero, shown here as the vertical line.

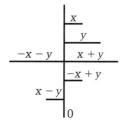

The diagram shows that
when x and y are both positive or both negative, $\quad |x + y| = |x| + |y|$
but when one is positive and the other is negative, $\quad |x + y| < |x| + |y|$,
so $|x + y| \leqslant |x| + |y|$

The modulus of a function

The graph of any curve whose equation is $y = |f(x)|$ can be found from the curve C with equation $y = f(x)$, by reflecting in the x-axis the parts of C for which $f(x)$ is negative. The remaining sections are not changed.

For example, to sketch $y = |(x - 1)(x - 2)|$ we start by sketching the curve $y = (x - 1)(x - 2)$

We then reflect in the x-axis the part of this curve which is below the x-axis.

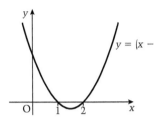
$y = (x - 1)(x - 2)$

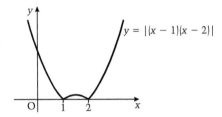
$y = |(x - 1)(x - 2)|$

For any function f, the mapping $x \mapsto |f(x)|$ is also a function.

Example

Sketch the graph of $y = |1 - \frac{1}{2}x|$ and write the equations in non-modulus form of each part on the sketch.

Start with a sketch of $y = 1 - \frac{1}{2}x$, then reflect the part below the x-axis in the x-axis.

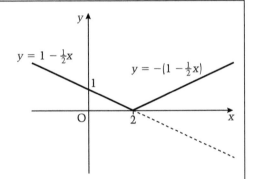
$y = 1 - \frac{1}{2}x$

$y = -(1 - \frac{1}{2}x)$

Example

Sketch the graph of $y = 2 + |x^2 - 4|$

Start with $y = |x^2 - 4|$, then translate the curve by 2 units in the positive direction of the y-axis.

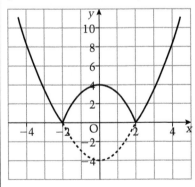

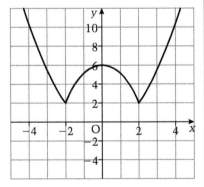

Exercise 1.21

Sketch the graphs of:

1 $y = 1 + |x - 1|$ **2** $y = \left|\dfrac{1}{x + 2}\right|$ **3** $y = |(x - 1)(x + 1)(x - 2)|$

Learning outcomes

- To solve modulus equations and inequalities

You need to know

- How to sketch the graph of $y = |f(x)|$
- How to solve linear and quadratic equations
- How to solve inequalities
- The properties of modulus functions

Intersection

To find the points of intersection between two graphs, we need to solve the equations of the graphs simultaneously. When those equations involve a modulus, a sketch helps to identify those equations in non-modulus form.

For example, to find the values of x where the graph of $y = |x + 2|$ intersects the graph $y = |1 - 2x|$, we draw a sketch and write on it the equations of each section in non-modulus form.

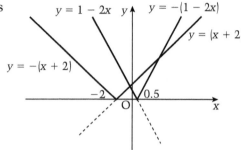

There are two points of intersection, one where

$$-(1 - 2x) = x + 2 \Rightarrow x = 3$$

and one where $1 - 2x = x + 2 \Rightarrow x = -\frac{1}{3}$

Alternatively, using the property that $|x| = |y| \Leftrightarrow x^2 = y^2$:

$$|x + 2| = |1 - 2x| \Rightarrow (x + 2)^2 = (1 - 2x)^2$$

$$\therefore \qquad x^2 + 4x + 4 = 4x^2 - 4x + 1$$

$$\Rightarrow 3x^2 - 8x - 3 = 0$$

$$\Rightarrow (3x + 1)(x - 3) = 0$$

$$\therefore \quad x = -\frac{1}{3} \text{ or } x = 3$$

Check:

when $x = -\frac{1}{3}$, $|x + 2| = 1\frac{2}{3}$ and $|1 - 2x| = 1\frac{2}{3}$, so $x = -\frac{1}{3}$ is a solution

when $x = 3$, $|x + 2| = 5$ and $|1 - 2x| = 5$, so $x = 3$ is also a solution.

It is essential that the values of x found using this method are checked because squaring can sometimes give values of x that are not solutions.

Solving equations involving modulus signs

We can solve an equation such as $|2x - 1| = 3x$ by sketching graphs as illustrated above, or by using the following fact:

when $|f(x)| = g(x)$ then $f(x) = g(x)$ and $-f(x) = g(x)$

Example

Solve the equation $|2x - 1| = 3x$

$2x - 1 = 3x$ gives $x = -1$ and $-(2x - 1) = 3x$ gives $x = \frac{1}{5}$

Check:

when $x = -1$, $|2x - 1| = 3$ and $3x = -3$, so $x = -1$ is not a solution

when $x = \frac{1}{5}$, $|2x - 1| = \frac{3}{5}$ and $3x = \frac{3}{5}$, so $x = \frac{1}{5}$ is a solution.

Therefore $x = \frac{1}{5}$ is the only solution.

Example

Solve the equation $x + 2 = \left| \dfrac{1}{x - 3} \right|$

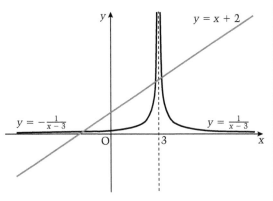

From the sketch, $x + 2 = \left| \dfrac{1}{x - 3} \right|$ where

$$x + 2 = \frac{1}{x - 3} \quad \Rightarrow x^2 - x - 7 = 0$$

$$\Rightarrow x = 3.19 \ (3 \text{ s.f.})$$

(the sketch shows that we only want the positive root)

and where

$$x + 2 = -\frac{1}{x - 3} \quad \Rightarrow x^2 - x - 5 = 0$$

$$\Rightarrow x = 2.79 \text{ or } -1.79 \ (3 \text{ s.f.})$$

Solving inequalities involving modulus signs

Simple inequalities can be solved from a sketch of the graphs.

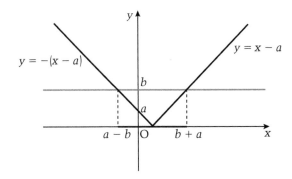

For example, the sketch shows that the inequality $|x - a| < b$ is true for $a - b < x < a + b$

Otherwise the method used is the same as for equations.

Example

Solve the inequality $|3 - x| < |x|$

From the sketch, $|3 - x| = |x|$ where $3 - x = x$,

i.e. where $x = \dfrac{3}{2}$

$\therefore |3 - x| < |x|$ for $x > \dfrac{3}{2}$

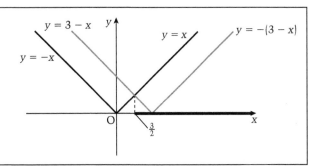

Exercise 1.22

1 Solve these equations:

  (a) $|3x - 2| = 5$

  (b) $|2 - x| = |x|$

  (c) $|e^x - 2| = 1$

2 Solve the following inequalities:

  (a) $|2x - 1| > |1 - x|$

  (b) $\left| \dfrac{1}{x + 1} \right| < |x + 1|$

  (c) $|\ln x| > x - 1$

Section 1   Practice questions

1 $f(n) = n^3 + 2n$

   (a) Show that $f(k + 1) = f(k) + 3(k^2 + k + 1)$

   (b) Hence prove by induction that $f(n)$ is divisible by 3 for all positive integer values of n.

2 $f(n) = 9^n - 1$

   (a) Show that $f(k + 1) = 9f(k) + 8$

   (b) Hence prove by induction that $f(n)$ is divisible by 8 for all positive integer values of n.

3 Prove by induction that $n^3 - n$ is a multiple of 6 for all positive integer values of n.

4 p and q are propositions.

   Construct a truth table to show the truth values of $\sim p \to q$ and $p \lor q$

   Hence determine whether $\sim p \to q$ and $p \lor q$ are equivalent statements.

5 p and q are propositions.

   (a) Write down the contrapositive of $\sim p \to \sim q$

   (b) Use the algebra of propositions to show that $\sim p \to (p \land \sim q) = q \to p$

6 (a) Sketch the graphs of
$$y = x + 1 \text{ and } y = \left| \frac{2}{x} \right|$$
on the same diagram. Show the coordinates of the points where the graphs intersect.

   (b) Find the range of values of x for which
$$x + 1 < \left| \frac{2}{x} \right|$$

7 Prove by induction that
$$1^2 + 2^2 + 3^2 + \ldots + n^2 = \frac{n}{6}(n + 1)(2n + 1)$$

8 The binary operation $\star$ is defined by
$x \star y = x^2 + y^2$
for all x and $y \in \mathbb{R}$.

   State with reasons whether the operation $\star$ is:
   (a) closed (b) commutative.

9 $f(x) = 2x^3 + 5x^2 + px + q$.
When $f(x)$ is divided by $(x + 2)$ the remainder is -10.
One root of the equation $f(x) = 0$ is $\frac{1}{2}$.

   (a) Find the values of p and q.

   (b) Factorise $f(x)$ and hence state the number of real roots of the equation $f(x) = 0$

10 (a) Solve the equation $e^x - 4e^{-x} - 3 = 0$

   (b) Find the values of x for which
$$\log_2 x - 2\log_x 2 = 1$$

11 $(x + 1)$ and $(x - 2)$ are factors of
$x^4 + px^3 + qx^2 - 16x - 12$

   (a) Find the values of p and q.

   (b) Hence solve the equation
$$x^4 + px^3 + qx^2 - 16x - 12 = 0$$

12 Given that $y = \dfrac{x^2}{x - 1}$ for all real values of x, find the range of values of y.

13 (a) Simplify $\dfrac{\sqrt{18} - 1}{\sqrt{2} + 3}$

   (b) Given that $x = 5^3$, find the value of $\log_x 5$.

14 Find the range of values of x for which
$$\frac{x - 1}{x^2 + 5x + 6} > 0.$$

15 (a) Sketch the graph of $y = \dfrac{2x}{x + 1}$.

   (b) On a separate diagram sketch the graph of
$$y = \left| \frac{2x}{x + 1} \right|.$$

   (c) Solve the equation $x = \left| \dfrac{2x}{x + 1} \right|$.

16 The binary operation $\star$ is defined by
$x \star y = x^2 - y^2$ for $x, y \in \mathbb{R}$.
Explain with reasons whether the operation is:
   (a) closed
   (b) associative
   (c) distributive over multiplication
   (d) such that x has an inverse.

17 (a) Sketch the graph of $y = 2 - |1 - x^2|$

   (b) Solve the inequality $|\ln x| > 2$

18 Given $f(x) = 2 - \ln x$, sketch the graph of:
   (a) $y = f(x)$
   (b) $y = f^{-1}(x)$

19 Solve the equation $6^{x - 1} = 5(3^{2x + 1})$

20 $f(x) = (x - 1)^2$ for $x \in \mathbb{R}$ and
$g(x) = \dfrac{1}{(x - 1)}$ for $x \neq 0, x \in \mathbb{R}$.

   (a) Find $fg(x)$ and $gf(x)$.

   (b) Explain why $g(x)$ has an inverse but $f(x)$ does not.

(c) Find $g^{-1}f(x)$.

(d) Define a domain for the function h where $h(x) = (x - 1)^2$ so that h^{-1} exists.

21 The function f is given by
$f = \{(2, 3), (3, 2), (4, 4), (1, 2)\}$

(a) Show that f is onto but not one-to-one.

(b) Suggest a change to one of the ordered pairs of f to give a function g such that g is both onto and one-to-one.

(c) Using your definition of g, give g^{-1} as a set of ordered pairs.

22 Find the relationship between a and b such that the line $x + ay + b = 0$ is a tangent to the curve $y = 2x^2 + x - 4$

Hence find the coordinates of the point where the tangent with gradient 1 touches the curve.

23 Find the maximum value of k for which
$$\frac{x}{x^2 + 1} \geq k - (x + 1)^2$$
for all real values of x.

24 Given that $x^2 + 1$ is a factor of
$x^4 + 3x^3 - 3x^2 + 3x - 4$,
find the other factor and hence factorise
$x^4 + 3x^3 + x^2 + 3x + 2$ completely.

25 On the same diagram, sketch the following curves:

(a) $y = x^2, y = (x + 1)^2, y = x^2 + 1$,
for $-2 < x < 2$,

(b) $y = \ln x, y = 1 + \ln x, y = 1 - \ln x$,
for $0 < x < 2$,

(c) $y = e^x, y = 3e^x, y = e^{3x}$, for $-2 < x < 2$

26 The roots of the equation
$2x^3 - 5x^2 + 6x + 3 = 0$
are α, β and γ.
Find the equation whose roots are $\alpha - 1, \beta - 1$ and $\gamma - 1$.

27 Factorise completely:

(a) $81x^4 - 16$

(b) $(a + b)^3 - b^3$

28 The function f is defined by
$$f(x) = \begin{cases} x - 2, x \leq 2 \\ 4 - x, x > 4 \end{cases} \text{ for } x \in \mathbb{R}.$$

(a) Find $ff(2)$.

(b) Explain why f does not have an inverse.

29 Show that $e^{\ln x} = x$.
Hence find, in terms of e, the value of
$e^{(\ln 6 + 2\ln 3 - \ln 13)}$.

30 Find the range of f where f is defined by
$$f(x) = \frac{x^2}{x + 1} \text{ for all } x \in \mathbb{R}.$$
Hence sketch the curve whose equation is
$y = f(x)$

31 p and q are propositions.
Draw a truth table for $\sim p \vee q$.

32 (a) Give a counter example to show that the following statement is false.
The sum of any two prime numbers is an even number.

(b) Prove that if n is any integer, $n^2 + n$ is an even integer.

33 Simplify $\dfrac{x - \sqrt{y}}{x + \sqrt{y}} + \dfrac{x + \sqrt{y}}{x - \sqrt{y}}$

34 Find the conditions satisfied by a and b such that $(x - 2)$ is a factor of $(x - a)(x^2 - 3b + 2b^2)$

35 Solve the inequality $|x^4 - 16| < 1$.

36 The function f is defined by
$$f(x) = \begin{cases} x^2 + 2, & x > 1 \\ x + 2, & x \leq 1 \end{cases}$$

(a) Sketch the graph of $y = f(x)$.

(b) Find **(i)** $f(0)$ **(ii)** $f(-2)$

(c) $g(x) = x^2 - 3x - 10$
Find the points of intersection of the curves
$y = f(x)$ and $y = g(x)$.

2.1 Sine, cosine and tangent functions

Learning outcomes

- To revise circular measure
- To revise the sine, cosine and tangent functions

You need to know

- The sine, cosine and tangent of an angle in a right-angled triangle
- The exact values of the sine, cosine and tangent of 30°, 45° and 60°
- How to use transformations of curves to help with curve sketching

The definition of an angle

When a line rotates from its initial position OP_0 about the fixed point O to any other position OP, the amount of rotation is measured by the angle between OP_0 and OP.

When the rotation is anticlockwise, the angle is positive, and when the rotation is clockwise, the angle is negative,

i.e. **a negative angle represents a clockwise rotation**.

The rotation of OP can be more than one revolution.

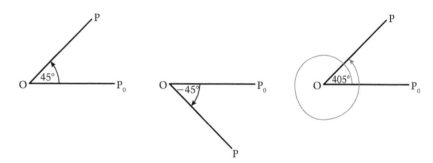

The radian

Degrees and revolutions are two units used to measure angle. The **radian** (sometimes called circular measure) is another unit used to measure angles.

One radian is the angle subtended at the centre of a circle by an arc equal in length to the radius of the circle.

The circumference of a circle is $2\pi r$ so the number of radians in one revolution is $\dfrac{2\pi r}{r} = 2\pi$ Therefore 2π rad $= 360°$.

The diagrams show some other angles measured in radians.

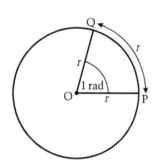

The sine, cosine and tangent functions

When OP is drawn on x- and y-axes, where $OP = 1$ unit and the coordinates of P are (x, y), then the sine, cosine and tangent functions are defined as:

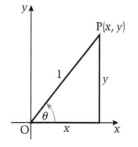

$$\sin\theta = \frac{y}{1}, \quad \cos\theta = \frac{x}{1} \quad \text{and} \quad \tan\theta = \frac{y}{x} \quad \text{for } \theta \in \mathbb{R}$$

The sine function

Measuring θ in radians, for $0 \leqslant \theta \leqslant \frac{\pi}{2}$, OP is in the first quadrant, y is positive and increases in value from 0 to 1 as θ increases from 0 to $\frac{\pi}{2}$.
Therefore $\sin \theta$ increases from 0 to 1.

For $\frac{\pi}{2} \leqslant \theta \leqslant \pi$, OP is in the second quadrant, y is positive and decreases in value from 1 to 0 as θ increases from $\frac{\pi}{2}$ to π.
Therefore $\sin \theta$ decreases from 1 to 0.

For $\pi \leqslant \theta \leqslant \frac{3\pi}{2}$, OP is in the third quadrant, y is negative and decreases in value from 0 to -1 as θ increases from π to $\frac{3\pi}{2}$.
Therefore $\sin \theta$ decreases from 0 to -1.

For $\frac{3\pi}{2} \leqslant \theta \leqslant 2\pi$, OP is in the fourth quadrant, y is negative and increases in value from -1 to 0 as θ increases from $\frac{3\pi}{2}$ to 2π.
Therefore $\sin \theta$ increases from -1 to 0.

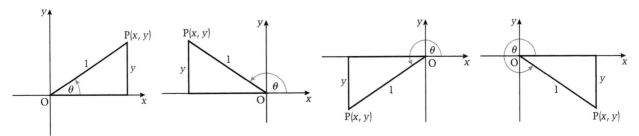

This shows that $\sin \theta$ is positive for $0 < \theta < \pi$ and negative for $\pi \leqslant \theta \leqslant 2\pi$ and $\sin \theta$ varies in value between -1 and 1. The pattern repeats itself as OP moves round the quadrants again.

As OP rotates clockwise, $\sin \theta$ decreases from 0 to -1, then increases from -1 to 0, and so on. The graph of $f(\theta) = \sin \theta$ shows these properties:

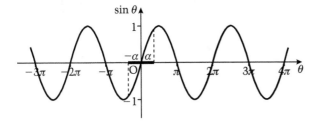

This graph also shows that, for any angle θ, $\sin(-\theta) = -\sin \theta$

The cosine function

For any value of θ, $\cos \theta = \frac{x}{1}$.

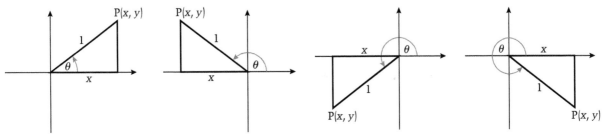

As OP moves round the quadrants, $\cos\theta$ decreases from 1 to 0, and then decreases again to -1, then increases to 0 and in the fourth quadrant increases again to 1. So $\cos\theta$, like $\sin\theta$, varies in value between -1 and 1.

The graph of $y = \cos\theta$ looks like this:

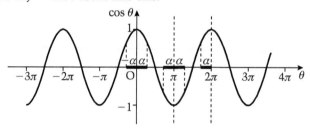

The curve is symmetrical about the vertical axis, showing that
$$\cos(-\theta) = \cos\theta$$

Comparing the curves for $\cos\theta$ and $\sin\theta$, we see that when we translate the cosine curve by $\frac{\pi}{2}$ in the positive direction of the horizontal axis, we get the sine curve, i.e. $\sin\theta = \cos\left(\theta - \frac{\pi}{2}\right)$

The tangent function

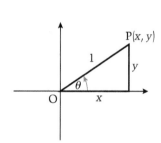

For any value of θ, $\tan\theta = \frac{y}{x}$

As OP rotates through the first quadrant, x decreases from 1 to 0 and y increases from 0 to 1. Therefore the fraction $\frac{y}{x}$ increases from 0 to very large values, and as $\theta \to \frac{\pi}{2}$, $\tan\theta \to \infty$

The behaviour of $\frac{y}{x}$ in the other quadrants shows that in the second quadrant $\tan\theta$ is negative and increases from $-\infty$ to 0, in the third quadrant $\tan\theta$ is positive and increases from 0 to ∞, and in the fourth quadrant $\tan\theta$ is negative and increases from $-\infty$ to 0
The cycle then repeats itself.

Therefore the graph of $f(\theta) = \tan\theta$ looks like this:

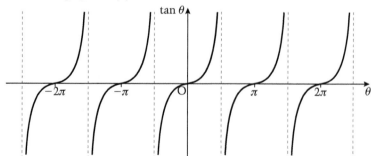

Properties of the sine, cosine and tangent functions

The graphs of the sine, cosine and tangent functions show that:

For $f(\theta) = \sin\theta$, $\theta \in \mathbb{R}$ $-1 \leqslant \sin\theta \leqslant 1$
$$\sin(-\theta) = -\sin\theta$$

      $\sin\theta$ is continuous (i.e. has no breaks) and has a pattern that repeats every 2π rad

For $f(\theta) = \cos\theta$, $\theta \in \mathbb{R}$ $-1 \leqslant \cos\theta \leqslant 1$
$$\cos(-\theta) = \cos\theta$$

      $\cos\theta$ is continuous (i.e. has no breaks) and has a pattern that repeats every 2π rad

For $f(\theta) = \tan \theta$, $\theta \in \mathbb{R}$ the range is unlimited

$$\tan(-\theta) = -\tan \theta$$

$\tan \theta$ is undefined when $\theta = \ldots -\dfrac{\pi}{2}, \dfrac{\pi}{2}, \dfrac{3\pi}{2}, \ldots$

and has a pattern that repeats every π rad.

Example

Find exact values for: **(a)** $\sin\dfrac{5\pi}{3}$ **(b)** $\cos\dfrac{5\pi}{3}$ **(c)** $\tan\dfrac{5\pi}{3}$

(a) From the sketch,

$$\sin\frac{5\pi}{3} = -\sin\frac{\pi}{3} = -\frac{\sqrt{3}}{2}$$

(b) From the sketch,

$$\cos\frac{5\pi}{3} = \cos\frac{\pi}{3} = \frac{1}{2}$$

(c) From the sketch,

$$\tan\frac{5\pi}{3} = -\tan\frac{\pi}{3} = -\sqrt{3}$$

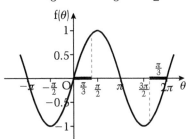
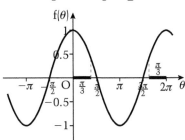
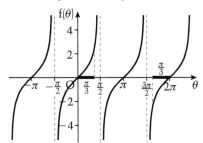

Example

Sketch the graphs of the following functions for values of θ in the range $0 \leqslant \theta \leqslant 2\pi$. In each case give the maximum value of the function.

(a) $f(\theta) = \sin 2\theta$ **(b)** $f(\theta) = 2 - 3\cos\theta$ **(c)** $f(\theta) = 2\sin\left(\dfrac{\theta}{2} - \dfrac{\pi}{2}\right)$

(a) Comparing $y = \sin 2\theta$ with $y = f(ax)$ shows that $y = \sin 2\theta$ is compressed by a factor of 2 parallel to the horizontal axis. Therefore $\sin 2\theta$ goes through two cycles for every one cycle of $\sin \theta$.

(b) Start with $y = \cos\theta$, then reflect the curve in the horizontal axis, stretch this curve by a factor of 3 parallel to the vertical axis and then translate the curve by 2 units up the vertical axis.

(c) Start with $y = \sin\dfrac{\theta}{2}$ (this is $\sin\theta$ stretched by a factor of 2 parallel to the horizontal axis) then translate by π in the positive direction of the horizontal axis, and lastly stretch by a factor of 2 parallel to the vertical axis.

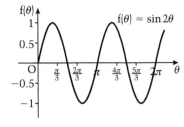

The maximum value is 1.

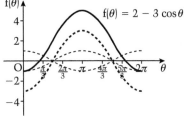

The maximum value is 5.

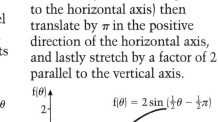
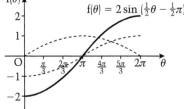

The maximum value is 2.

Exercise 2.1

1 Find exact values for: **(a)** $\cos\left(-\dfrac{5\pi}{4}\right)$ **(b)** $\tan\left(\dfrac{7\pi}{4}\right)$ **(c)** $\sin\left(\dfrac{9\pi}{2}\right)$

2 Sketch the graphs of the following functions for values of θ in the range $0 \leqslant \theta \leqslant 2\pi$

   (a) $f(\theta) = \tan 2\theta$ **(b)** $f(\theta) = 2 - \cos\theta$ **(c)** $f(\theta) = \sin\left(\theta + \dfrac{\pi}{2}\right)$

3 Write down the maximum and minimum values of each function.

   (a) $5\cos(2\theta - \pi)$ **(b)** $5 + \cos 2\theta$ **(c)** $\dfrac{1}{2} - 2\sin\theta$

- To define and use the reciprocal trigonometric functions

- The properties and graphs of the sine, cosine and tangent functions

The reciprocal trigonometric functions

The reciprocals of the three main trigonometric (trig) functions have their own names:

$$\frac{1}{\sin \theta} \equiv \operatorname{cosec} \theta \qquad \frac{1}{\cos \theta} \equiv \sec \theta \qquad \frac{1}{\tan \theta} \equiv \cot \theta$$

where cosec is the abbreviation of cosecant, sec is the abbreviation for secant and cot is the abbreviation for cotangent.

The cosecant function

The graph of $f(\theta) = \operatorname{cosec} \theta$ is given below.

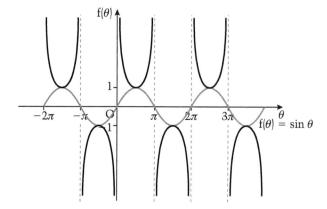

The graph shows that:

- the cosec function is not continuous; it is undefined when θ is any multiple of π (this is to be expected because these are the values of θ where $\sin \theta = 0$ and $\frac{1}{0}$ is undefined)

- the cosec function takes all real values except for values between -1 and 1.

The secant function

The graph of $f(\theta) = \sec \theta$, shown below, is similar to the graph for cosec θ.

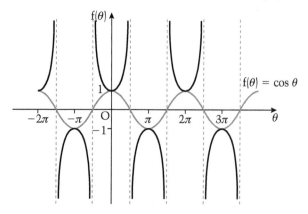

The properties of the secant function are similar to those of the cosecant function:

- it is not continuous; it is undefined when θ is any odd multiple of $\frac{\pi}{2}$
- it takes all real values except for values between -1 and 1.

The cotangent function

The graph of $f(\theta) = \cot \theta$ is given below.

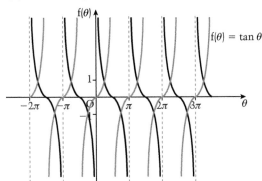

The properties of the cotangent function are similar to those of the tangent function:

- it is undefined when θ is any multiple of π
- it takes all real values.

From the graph we can see that the curve for $\cot \theta$ is the reflection of the curve for $\tan \theta$ in the x-axis, and it is translated by $\frac{\pi}{2}$ in the direction of the negative x-axis.

Therefore $\qquad \cot \theta = \tan \left(-\theta + \frac{\pi}{2}\right) = \tan \left(\frac{\pi}{2} - \theta\right)$

Example

Find, in radians, the smallest value of θ for which $\sec \theta = -2$

$\cos \theta = \dfrac{1}{\sec \theta} = -\dfrac{1}{2}$

From the sketch, the required value of θ is $\dfrac{2\pi}{3}$

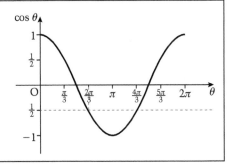

Example

Find the exact value of $\operatorname{cosec} \dfrac{7\pi}{6}$

$\operatorname{cosec} \dfrac{7\pi}{6} = \dfrac{1}{\sin \frac{7\pi}{6}}$

From the sketch, $\sin \dfrac{7\pi}{6} = -\dfrac{1}{2}$

$\operatorname{cosec} \dfrac{7\pi}{6} = -2$

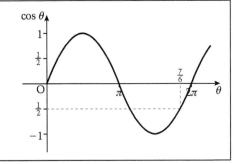

Exercise 2.2

1. Find the exact value of: **(a)** $\cot \dfrac{\pi}{4}$ **(b)** $\sec \dfrac{7\pi}{4}$ **(c)** $\operatorname{cosec} \dfrac{5\pi}{3}$

2. Find, in radians, the smallest value of x for which:
   (a) $\cot x = -1$ **(b)** $\sec x = -\dfrac{2}{\sqrt{3}}$ **(c)** $\operatorname{cosec} x = -1$

Learning outcomes

- To derive and use the Pythagorean identities

You need to know

- The definitions of the sine, cosine and tangent functions
- The definitions of the reciprocal trig functions
- Pythagoras' theorem
- How to solve a quadratic equation

The relationship between the sine, cosine and tangent of any angle

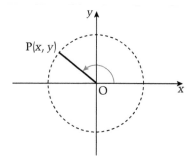

For any angle θ, we know that $\sin \theta = \dfrac{y}{OP}$, $\cos \theta = \dfrac{x}{OP}$, $\tan \theta = \dfrac{y}{x}$

But $\dfrac{\sin \theta}{\cos \theta} = \dfrac{y}{OP} \Big/ \dfrac{x}{OP} = \dfrac{y}{x}$

i.e. for all values of θ, $\tan \theta \equiv \dfrac{\sin \theta}{\cos \theta}$　　　　　　[1]

Using the reciprocal functions, this identity can be written as

$$\cot \theta \equiv \dfrac{\cos \theta}{\sin \theta}$$

The Pythagorean identities

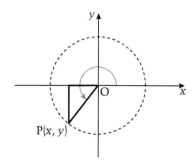

For any angle θ, a right-angled triangle can be drawn in which $x^2 + y^2 = OP^2$

Dividing by OP^2 gives $\left(\dfrac{x}{OP}\right)^2 + \left(\dfrac{y}{OP}\right)^2 = 1$

i.e. $\cos^2 \theta + \sin^2 \theta \equiv 1$　　　　　　[2]
($\cos^2 \theta$ means $(\cos \theta)^2$, etc.)

We can use these identities to produce further identities.

Dividing $\cos^2 \theta + \sin^2 \theta \equiv 1$ by $\cos^2 \theta$ gives $1 + \dfrac{\sin^2 \theta}{\cos^2 \theta} \equiv \dfrac{1}{\cos^2 \theta}$

then using $\tan \theta \equiv \dfrac{\sin \theta}{\cos \theta}$ gives 　　　　**$1 + \tan^2 \theta \equiv \sec^2 \theta$**　　[3]

Dividing $\cos^2 \theta + \sin^2 \theta \equiv 1$ by $\sin^2 \theta$ gives $\dfrac{\cos^2 \theta}{\sin^2 \theta} + 1 \equiv \dfrac{1}{\sin^2 \theta}$

then using $\cot \theta \equiv \dfrac{\cos \theta}{\sin \theta}$ gives 　　　　**$\cot^2 \theta + 1 \equiv \mathrm{cosec}^2 \theta$**　[4]

These identities can also be used to solve some trig equations.

Example

Solve the equation $2\cos^2 x - \sin x = 1$ for $0 \leqslant x \leqslant 2\pi$

Using [2] gives $\cos^2 x = 1 - \sin^2 x$ so $2\cos^2 x - \sin x = 1$ becomes

$$2(1 - \sin^2 x) - \sin x = 1$$

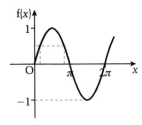

$\Rightarrow \qquad 2\sin^2 x + \sin x - 1 = 0$

This is a quadratic equation in $\sin x$

$\therefore \qquad (2\sin x - 1)(\sin x + 1) = 0$

$\Rightarrow \qquad \sin x = \frac{1}{2}$ or $\sin x = -1$

$\Rightarrow \qquad x = \dfrac{\pi}{6}, \dfrac{5\pi}{6}, \dfrac{3\pi}{2}$

These identities can be used to prove the validity of some other identities.

Example

Prove that $(1 - \cos A)(1 + \sec A) \equiv \sin A \tan A$

This identity has yet to be proved so do not use the complete identity in your working. Work separately on the left-hand side (LHS) and the right-hand side (RHS).

$$\text{LHS} = (1 - \cos A)(1 + \sec A) = 1 - \cos A + \sec A - \cos A \sec A$$

$$= 1 - \cos A + \sec A - \frac{\cos A}{\cos A}$$

$$= \sec A - \cos A$$

$$= \frac{1 - \cos^2 A}{\cos A} = \frac{\sin^2 A}{\cos A} = \sin A \frac{\sin A}{\cos A}$$

$$= \sin A \tan A = \text{RHS}$$

$\therefore \qquad (1 - \cos A)(1 + \sec A) \equiv \sin A \tan A$

These identities can also be used to eliminate trig ratios from a set of equations.

Example

Eliminate θ from the equations $x = 3\sin\theta$ and $y = 2\sec\theta$

$x = 3\sin\theta \Rightarrow \sin\theta = \dfrac{x}{3}$ and $y = 2\sec\theta \Rightarrow \cos\theta = \dfrac{2}{y}$

$\therefore \quad \left(\dfrac{x}{3}\right)^2 + \left(\dfrac{2}{y}\right)^2 = 1$ Using $\cos^2\theta + \sin^2\theta \equiv 1$

Exercise 2.3

1 Solve the equation $\sec^2 x + \tan^2 x = 3$ for $0 \leqslant x \leqslant 2\pi$

2 Simplify $\dfrac{\sin x}{\cos x} - \dfrac{3\cos x}{\sin x}$

3 Eliminate θ from the equations $x = 4\tan\theta$ and $y = 2\cos\theta$

4 Prove the identity $\tan A + \cot A \equiv \sec A \operatorname{cosec} A$

The identity $\cos(A - B) \equiv \cos A \cos B + \sin A \sin B$

Trigonometric functions are not distributive,
i.e. $\cos(A - B)$ is NOT equal to $\cos A - \cos B$

We can derive the correct identity for $\cos(A - B)$ using the diagram below, which shows a circle of radius 1 unit centre O.

We find the length of PQ using two different methods and then equate the two results.

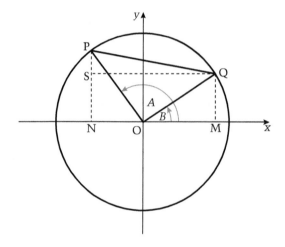

Using the cosine formula in $\triangle OPQ$ gives

$$PQ^2 = 1^2 + 1^2 - 2\cos(A - B)$$
$$= 2 - 2\cos(A - B)$$

From the diagram

$$OM = \cos B \quad \text{and} \quad ON = -\cos A \quad \text{so} \quad QS = (-\cos A + \cos B)$$
$$QM = \sin B \quad \text{and} \quad PN = \sin A \quad \text{so} \quad PS = (\sin A - \sin B)$$

Using Pythagoras' theorem in $\triangle PQS$ gives

$$PQ^2 = (-\cos A + \cos B)^2 + (\sin A - \sin B)^2$$
$$= \cos^2 A + \cos^2 B - 2\cos A \cos B + \sin^2 A + \sin^2 B - 2\sin A \sin B$$
$$= 2 - 2(\cos A \cos B + \sin A \sin B)$$

Equating the two expressions for PQ^2 gives

$$2 - 2\cos(A - B) = 2 - 2(\cos A \cos B + \sin A \sin B)$$
$$\Rightarrow \cos(A - B) \equiv \cos A \cos B + \sin A \sin B$$

The angles A and B can be any size and the proof is similar. Therefore the identity is true for all angles.

Compound angle identities

The identity derived above is one of the compound angle identities and
we can use it to derive others.

$$\cos(A - B) \equiv \cos A \cos B + \sin A \sin B \qquad [1]$$

Replacing B by $-B$ in [1], $\qquad \cos(A + B) \equiv \cos A \cos(-B) + \sin A \sin(-B)$

$$\equiv \cos A \cos B - \sin A \sin B \qquad [2]$$

Replacing A by $\frac{\pi}{2} - A$ in [1] gives

$$\cos\left(\frac{\pi}{2} - (A + B)\right) \equiv \cos\left(\frac{\pi}{2} - A\right)\cos B + \sin\left(\frac{\pi}{2} - A\right)\sin B$$

$\Rightarrow \qquad\qquad\qquad \sin(A + B) \equiv \sin A \cos B - \cos A \sin B \qquad [3]$

Replacing B by $- B$ in [3], $\qquad \sin(A - B) \equiv \sin A \cos(-B) + \sin A \cos(-B)$

$$\equiv \sin A \cos B - \cos A \sin B \qquad [4]$$

Dividing [1] by [3] gives $\qquad \dfrac{\sin(A + B)}{\cos(A + B)} \equiv \dfrac{\sin A \cos B + \cos A \sin B}{\cos A \cos B - \sin A \sin B}$

$$\tan(A + B) \equiv \dfrac{\dfrac{\sin A \cos B}{\cos A \cos B} + \dfrac{\cos A \sin B}{\cos A \cos B}}{\dfrac{\cos A \cos B}{\cos A \cos B} - \dfrac{\sin A \sin B}{\cos A \cos B}}$$

$\Rightarrow \qquad\qquad\qquad\qquad \equiv \dfrac{\tan A - \tan B}{1 - \tan A \tan B} \qquad [5]$

Replacing B by $-B$ in [5] gives

$$\tan(A - B) \equiv \dfrac{\tan A - \tan B}{1 + \tan A \tan B} \qquad [6]$$

**You need to learn these identities and be able to
recognise either side.**

Collecting the identities together gives

$$\sin(A + B) \equiv \sin A \cos B + \cos A \sin B$$

$$\sin(A - B) \equiv \sin A \cos B - \cos A \sin B$$

$$\cos(A + B) \equiv \cos A \cos B - \sin A \sin B$$

$$\cos(A - B) \equiv \cos A \cos B + \sin A \sin B$$

$$\tan(A + B) = \dfrac{\tan A + \tan B}{1 - \tan A \tan B}$$

$$\tan(A - B) = \dfrac{\tan A - \tan B}{1 + \tan A \tan B}$$

Using compound angle identities

These identities can be used to prove further identities.

Example

Prove that $\sin(A + \pi) + \sin(A - \pi) \equiv -2\sin A$

$$\text{LHS} = \sin A \cos \pi + \cos A \sin \pi + \sin A \cos \pi - \cos A \sin \pi$$

$$= 2\sin A \cos \pi$$

$$\cos \pi = -1$$

Therefore $\text{LHS} = -2\sin A = \text{RHS}$

These identities can be used to solve equations.

☑ *Exam tip*

The range is given in radians so the answer must be given in radians.

Example

Solve the equation $\cos \theta = \sin\left(\theta - \frac{\pi}{3}\right)$ for values of θ in the range $0 \leq \theta \leq \pi$

$$\cos \theta = \sin\left(\theta - \frac{\pi}{3}\right)$$

$$= \sin \theta \cos \frac{\pi}{3} - \cos \theta \sin \frac{\pi}{3}$$

$$= \frac{1}{2}\sin \theta - \frac{\sqrt{3}}{2}\cos \theta$$

$$\therefore \quad \left(1 + \frac{\sqrt{3}}{2}\right)\cos \theta = \frac{1}{2}\sin \theta$$

$$\Rightarrow (2 + \sqrt{3})\cos \theta = \sin \theta$$

$$\Rightarrow \tan \theta = 2 + \sqrt{3}$$

$$\therefore \quad \theta = 1.31 \text{ rad (3 s.f.)}$$

The identities can be used to find exact values of some trig ratios.

Example

Find the exact value of $\sin 15°$.

$$\sin 15° = \sin(45° - 30°) = \sin 45° \cos 30° - \cos 45° \sin 30°$$

$$= \frac{1}{\sqrt{2}} \times \frac{\sqrt{3}}{2} - \frac{1}{\sqrt{2}} \times \frac{1}{2}$$

$$= \frac{\sqrt{3}}{2\sqrt{2}} - \frac{1}{2\sqrt{2}}$$

$$= \frac{\sqrt{6} - \sqrt{2}}{4}$$

They can be used to simplify expressions.

Example

Simplify $\cos \theta \cos 2\theta + \sin \theta \sin 2\theta$

This is the RHS of [2] with A $= \theta$ and B $= 2\theta$

$$\cos \theta \cos 2\theta + \sin \theta \sin 2\theta = \cos(\theta - 2\theta)$$

$$= \cos(-\theta)$$

$$= \cos \theta$$

These identities can also be used to eliminate an angle from two equations.

Example

Eliminate θ from the equations $x = \sin\left(\theta + \frac{\pi}{4}\right)$ and $y = \cos\theta$

$$x = \sin\left(\theta + \frac{\pi}{4}\right)$$
$$= \sin\theta\cos\frac{\pi}{4} + \cos\theta\sin\frac{\pi}{4}$$
$$= \frac{1}{\sqrt{2}}\sin\theta + \frac{1}{\sqrt{2}}\cos\theta$$
$$= \frac{1}{\sqrt{2}}\sin\theta + \frac{1}{\sqrt{2}}y$$
$$x(\sqrt{2} - y) = \sin\theta$$

Then, using $\sin^2\theta + \cos^2\theta \equiv 1$ gives
$$(x\sqrt{2} - y)^2 + y^2 = 1$$
$$\Rightarrow 2x^2 + 2y^2 - 2xy\sqrt{2} = 1$$

Exercise 2.4

1 Find the exact value of:

(a) $\sin 75°$

(b) $\sin 50°\cos 40° + \cos 50°\sin 40°$

2 Prove that:

(a) $\cot(A + B) = \dfrac{\cot A \cot B - 1}{\cot A + \cot B}$

(b) $\sin(60° + \theta) = \sin(120° - \theta)$

3 Simplify:

(a) $\sin\theta\cos 3\theta + \cos\theta\sin 3\theta$

(b) $\dfrac{\tan P - \tan 3P}{1 + \tan P \tan 3P}$

4 Solve the equations for $0 \leqslant \theta \leqslant \pi$

(a) $\cos\left(\theta + \frac{\pi}{3}\right) = \sin\theta$

(b) $\cos\left(\frac{\pi}{4} - \theta\right) = \sin\theta$

5 Eliminate θ from the equations
$$x = 4\tan\left(\theta - \frac{\pi}{4}\right) \text{ and } y = 2\tan\theta$$

6 Given that $\sin\alpha = \frac{3}{5}$, find the exact value of:

(a) $\sin\left(\alpha + \frac{\pi}{4}\right)$

(b) $\tan\left(\alpha + \frac{\pi}{4}\right)$

2.5   Double angle identities

Double angle identities

The compound angle identities are true for any two angles, A and B, so they can be used for two equal angles, i.e. when $B = A$

Replacing B by A in the compound angle formulae gives

$$\sin 2A \equiv 2 \sin A \cos A$$

$$\cos 2A \equiv \cos^2 A - \sin^2 A$$

$$\tan 2A \equiv \frac{2\tan A}{1 - \tan^2 A}$$

The identity for $\cos 2A$ can be expressed in other forms because

$$\cos^2 A - \sin^2 A = \cos^2 A - (1 - \cos^2 A) = 2\cos^2 A - 1$$

and

$$\cos^2 A - \sin^2 A = (1 - \sin^2 A) - \sin^2 A = 1 - 2\sin^2 A$$

Therefore

$$\cos 2A \equiv \begin{cases} \cos^2 A - \sin^2 A \\ 2\cos^2 A - 1 \\ 1 - 2\sin^2 A \end{cases}$$

The last two identities above can be rearranged as

$$\cos^2 A = \tfrac{1}{2}(1 + \cos 2A)$$

$$\sin^2 A = \tfrac{1}{2}(1 - \cos 2A)$$

You also need to learn all the double angle identities including all the alternative identities involving $\cos 2A$. The identities in this set are some of the most useful for simplifying trigonometric expressions.

As with the previous identities, they can be used in a variety of problems. The following examples illustrate some of their uses.

Example

Given that θ is an acute angle and that $\sin \theta = \frac{3}{5}$, find the value of :

(a) $\cos 3\theta$ **(b)** $\tan 2\theta$

Given θ is acute and $\sin \theta = \frac{3}{5}$, then $\cos \theta = \frac{4}{5}$ and $\tan \theta = \frac{3}{4}$

(a) $\cos 3\theta = \cos(2\theta + \theta) = \cos 2\theta \cos \theta - \sin 2\theta \sin \theta$

$\qquad\qquad\qquad\qquad\qquad\qquad$ Using a compound angle identity

$$= (2\cos^2 \theta - 1)\cos \theta - 2\sin^2\theta \cos \theta$$

$$= \left[2\left(\frac{16}{25}\right) - 1\right] \times \frac{4}{5} - 2 \times \frac{9}{25} \times \frac{4}{5} = -\frac{44}{125}$$

(b) $\tan 2\theta = \dfrac{2\tan\theta}{1 - \tan^2\theta} = \dfrac{2 \times \frac{3}{4}}{1 - \frac{9}{16}} = \dfrac{24}{7}$

Example

Solve the equation $\cos 2x + 3\sin x = 2$ for $0 \leqslant x \leqslant \pi$

$$\cos 2x + 3\sin x = 2$$

$\Rightarrow (1 - 2\sin^2 x) + 3\sin x = 2$

$\Rightarrow 2\sin^2 x - 3\sin x + 1 = 0$

$\Rightarrow (2\sin x - 1)(\sin x - 1) = 0$

$\Rightarrow \sin x = \frac{1}{2}$ or 1, so $x = \frac{\pi}{6}, \frac{5\pi}{6}, \pi$

Using the identity $\cos 2x \equiv 1 - 2\sin^2 x$ so that the equation contains only terms in $\sin x$

Example

Prove the identity $\sin 3x \equiv 3\sin x - 4\sin^3 x$

$\text{LHS} = \sin 3x = \sin(2x + x)$

$\qquad = \sin 2x \cos x + \cos 2x \sin x$

$\qquad = 2\sin x \cos x \cos x + (1 - 2\sin^2 x)\sin x$

$\qquad = 2\sin x(1 - \sin^2 x) + \sin x - 2\sin^3 x$

$\qquad = 3\sin x - 4\sin^3 x = \text{RHS}$

Therefore $\sin 3x \equiv 3\sin x - 4\sin^3 x$

The identity for sin 3x is worth remembering.

Note that we used a compound angle identity and a Pythagorean identity as well as a double angle identity in this proof.

Example

Eliminate θ from the equations $x = \cos 2\theta$ and $y = \text{cosec}\,\theta$

$$y = \frac{1}{\sin\theta} \quad \text{and} \quad x = 1 - 2\sin^2\theta$$

$$\therefore \quad x = 1 - \frac{2}{y^2} \Rightarrow y^2(1 - x) = 2$$

Exercise 2.5

1 Given that $\cos\theta = \frac{12}{13}$ and that θ is acute, find the exact value of:

(a) $\sin 2\theta$

(b) $\cos\frac{\theta}{2}$ (Hint: use a double angle identity with $A = \frac{\theta}{2}$)

2 Prove the identity $\cos 2A = \frac{1 - \tan^2 A}{1 + \tan^2 A}$

3 Solve the equation $\tan x \tan 2x = 2$ for $0 \leqslant x \leqslant \pi$

4 Eliminate t from the equations $x = \cos 2t$ and $y = \sec 4t$

The factor formulae

The identities in this set are called the factor formulae because they
convert expressions such as $\sin A + \sin B$ into a product.

Starting with the compound angle formulae

$$\sin A \cos B + \cos A \sin B \equiv \sin (A + B)$$

$$\sin A \cos B - \cos A \sin B \equiv \sin (A - B)$$

adding gives $2 \sin A \cos B \equiv \sin (A + B) + \sin (A - B)$ [1]

subtracting gives $2 \cos A \sin B \equiv \sin (A + B) - \sin (A - B)$ [2]

Now, using the other two compound angle formulae

$$\cos A \cos B - \sin A \sin B \equiv \cos (A + B)$$

$$\cos A \cos B + \sin A \sin B \equiv \cos (A - B)$$

adding gives $2 \cos A \cos B \equiv \cos (A + B) + \cos (A - B)$ [3]

subtracting gives $-2 \sin A \sin B \equiv \cos (A + B) - \cos (A - B)$ [4]

The right-hand side of each of these formulae can be simplified by the
following substitutions.

$$\left. \begin{array}{c} P = A + B \\ Q = A - B \end{array} \right\} \quad \Rightarrow \quad \left\{ \begin{array}{l} A = \frac{1}{2}(P + Q) \\ B = \frac{1}{2}(P - Q) \end{array} \right.$$

Then $\sin P + \sin Q \equiv 2 \sin \frac{1}{2}(P + Q) \cos \frac{1}{2}(P - Q)$ [5]

           $\sin P - \sin Q \equiv 2 \cos \frac{1}{2}(P + Q) \sin \frac{1}{2}(P - Q)$ [6]

           $\cos P + \cos Q \equiv 2 \cos \frac{1}{2}(P + Q) \cos \frac{1}{2}(P - Q)$ [7]

           $\cos P - \cos Q \equiv -2 \sin \frac{1}{2}(P + Q) \sin \frac{1}{2}(P - Q)$ [8]

You may find it easier to remember the last group using words.
For example,

sum of sines = twice sine (half sum) × cosine (half difference)

The first group, [1]–[4], is used when we want to express a product as a
sum or difference. For example, using [1]

$$\sin 5x \cos 3x = \tfrac{1}{2}(\sin 4x + \sin x)$$

The second group, [5]–[8], is used when we want to express a sum or
difference as a product. For example, using [5]

$$\sin 5x + \sin 3x = 2 \sin 4x \cos x$$

As well as being useful for solving some trig equations and proving some
identities, the factor formulae are particularly useful for certain types of
calculus problems which we will consider in Section 3.

Example

Prove that $\dfrac{\cos 2x - \cos 2y}{\sin 2x + \sin 2y} \equiv \tan(y - x)$

$\text{LHS} = \dfrac{\cos 2x - \cos 2y}{\sin 2x + \sin 2y} = \dfrac{-2\sin(x + y)\sin(x - y)}{2\sin(x + y)\cos(x - y)}$ Using [8] and [5]

$\qquad\qquad\qquad\quad = \dfrac{-\sin(x - y)}{\cos(x - y)}$

$\qquad\qquad\qquad\quad = \dfrac{\sin(y - x)}{\cos(y - x)}$ Using $\sin A = -\sin(-A)$ and $\cos A = \cos(-A)$

$\qquad\qquad\qquad\quad = \tan(y - x) = \text{RHS}$

$\therefore \quad \dfrac{\cos 2x - \cos 2y}{\sin 2x + \sin 2y} \equiv \tan(y - x)$

Example

Solve the equation $\cos 2x + \cos 4x = 0$ for $0 \leqslant x \leqslant 2\pi$

$\cos 2x + \cos 4x = 0$

$\Rightarrow \quad 2\cos 3x \cos x = 0$ Using [7]

$\Rightarrow \quad \cos 3x = 0$ or $\cos x = 0$

$\therefore$ either $\cos 3x = 0 \Rightarrow 3x = \dfrac{\pi}{2}, \dfrac{3\pi}{2}, \dfrac{5\pi}{2}, \dfrac{7\pi}{2}, \dfrac{9\pi}{2}, \dfrac{11\pi}{2}$

For values of x between 0 and 2π, we need values of $3x$ between 0 and 6π, i.e. in 3 times the range for x

$$\Rightarrow x = \dfrac{\pi}{6}, \dfrac{\pi}{2}, \dfrac{5\pi}{6}, \dfrac{7\pi}{6}, \dfrac{3\pi}{2}, \dfrac{11\pi}{6}$$

or $\cos x = 0 \qquad \Rightarrow x = \dfrac{\pi}{2}$ or $\dfrac{3\pi}{2}$

$\therefore x = \dfrac{\pi}{6}, \dfrac{\pi}{2}, \dfrac{5\pi}{6}, \dfrac{7\pi}{6}, \dfrac{3\pi}{2}, \dfrac{11\pi}{6}$

This example includes a special case of an equation of the form $\cos ax = b$. To find values of x in a given range, we need to find values of ax in a times that range.

The same is true for equations of the form $\sin ax = b$ and $\tan ax = b$

Exercise 2.6

1   Simplify $\sin\left(x + \dfrac{\pi}{3}\right) - \sin\left(x - \dfrac{\pi}{3}\right)$

2   Prove the identity $\sin A + \sin 2A + \sin 3A \equiv \sin 2A(1 + 2\cos A)$

3   Solve the equation $\sin 3x = \sin 4x + \sin 2x$ for $0 \leqslant x \leqslant \pi$

The expression $a \cos \theta + b \sin \theta$

The expression $a \cos \theta + b \sin \theta$ can be reduced to a single term such as $r \cos (\theta - \alpha)$

We can find values for r and α by expanding $r \cos (\theta - \alpha)$ using a compound angle identity,

i.e. if $\qquad r \cos (\theta - \alpha) \equiv a \cos \theta + b \sin \theta$

then $\quad r (\cos \theta \cos \alpha + \sin \theta \sin \alpha) \equiv a \cos \theta + b \sin \theta$

Comparing the coefficients of $\cos \theta$ and $\sin \theta$ gives

$$r \cos \alpha = a \quad [1] \qquad \text{and} \qquad r \sin \alpha = b \quad [2]$$

$$[2] \div [1] \Rightarrow \tan \alpha = \frac{b}{a}$$

and $[1]^2 + [2]^2 \Rightarrow r^2(\cos^2 \alpha + \sin^2 \alpha) = a^2 + b^2$

$$\Rightarrow r = \sqrt{a^2 + b^2} \quad \text{Using } \cos^2 \alpha + \sin^2 \alpha \equiv 1$$

However, it is not sensible to rely on the formulae for r and α. You should always use the method to work out the result.

For example, to express $4 \cos \theta + 3 \sin \theta$ as $r \cos (\theta - \alpha)$, start with the expansion of $r \cos (\theta - \alpha)$, which gives

$$r (\cos \theta \cos \alpha + \sin \theta \sin \alpha) = 4 \cos \theta + 3 \sin \theta$$

Comparing the coefficients of $\cos \theta$ and $\sin \theta$ gives $r \cos \alpha = 4$ and $r \sin \alpha = 3$

Then $\tan \alpha = \frac{3}{4}$ and $r = \sqrt{3^2 + 4^2} = 5$

Therefore $\quad 4 \cos \theta + 3 \sin \theta = 5 \cos (\theta - \alpha)$ where $\tan \alpha = \frac{3}{4}$

We can also express $4 \cos \theta + 3 \sin \theta$ as $r \sin (\theta + \alpha)$.

This time start with the expansion of $r \sin (\theta + \alpha)$, which gives

$$r (\sin \theta \cos \alpha + \cos \theta \sin \alpha) = 4 \cos \theta + 3 \sin \theta$$

Comparing the coefficients of $\cos \theta$ and $\sin \theta$ gives $r \cos \alpha = 3$ and $r \sin \alpha = 4$

Then $\tan \alpha = \frac{4}{3}$ and $r = \sqrt{4^2 + 3^2} = 5$

Therefore $4 \cos \theta + 3 \sin \theta = 5 \sin (\theta + \alpha)$ where $\tan \alpha = \frac{4}{3}$

Exercise 2.7a

1 Express $5 \cos \theta + 12 \sin \theta$ in the form:
   (a) $r \cos (\theta - \alpha)$ (b) $r \sin (\theta + \alpha)$

2 Express $7 \cos \theta - 24 \sin \theta$ in the form:
   (a) $r \cos (\theta + \alpha)$ (b) $r \sin (\theta - \alpha)$

Finding maximum and minimum values of $a \cos \theta + b \sin \theta$

To find the maximum and minimum values of an expression of the form $a \cos \theta + b \sin \theta$ we express it as a single sine or cosine. We can then 'read' the maximum and minimum values.

For example, to find the maximum and minimum values of $3\sin x - 2\cos x$, we can express $3\sin x - 2\cos x$ as $r\sin(x - \alpha)$:

$$r(\sin x \cos\alpha - \cos x \sin\alpha) = 3\sin x - 2\cos x$$

Comparing coefficients of $\sin x$ and $\cos x$ gives $r\cos\alpha = 3$ and $r\sin\alpha = 2$

Therefore $r = \sqrt{2^2 + 3^2} = \sqrt{13}$ and $\tan\alpha = \frac{2}{3}$

So $3\sin x - 2\cos x = \sqrt{13}\sin(x - \alpha)$ We do not need to evaluate α

The maximum value of the sine of an angle is 1 and the minimum value is -1, therefore the maximum value of $3\sin x - 2\cos x$ is $\sqrt{13}$ and the minimum value is $-\sqrt{13}$.

To find the value of x at which the maximum and minimum values occur, we do need to evaluate α:

$\tan\alpha = \frac{2}{3} \Rightarrow \alpha = 0.588\,\text{rad}$ Make sure your calculator is in radian mode

$\sin(x - \alpha)$ is maximum when $(x - \alpha) = \frac{\pi}{2}$ and minimum when

$(x - \alpha) = \frac{3\pi}{2}$, i.e. when $x = \frac{\pi}{2} + 0.588\,\text{rad}$ and $x = \frac{3\pi}{2} + 0.588\,\text{rad}$

The equation $a\cos\theta + b\sin\theta = c$

Expressing $a\cos\theta + b\sin\theta$ as a single sine or cosine makes solving equations of the form $a\cos\theta + b\sin\theta = c$ straightforward.

For example, to solve $\cos x + \sqrt{3}\sin x = 1$, for $0 \le x \le 2\pi$,

we can express $\cos x + \sqrt{3}\sin x$ as $r\cos(x - \alpha)$,

i.e. $r\cos x \cos\alpha + r\sin x \sin\alpha = \cos x + \sqrt{3}\sin x$

$\Rightarrow r^2 = 4$ and $\tan\alpha = \sqrt{3}$ so $\alpha = \frac{\pi}{3}$

$\therefore \quad \cos x + \sqrt{3}\sin x = 1$ becomes $2\cos\left(x - \frac{\pi}{3}\right) = 1$

$\Rightarrow \qquad\qquad\qquad\qquad \cos\left(x - \frac{\pi}{3}\right) = \frac{1}{2}$

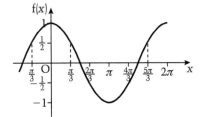

To find values of x in the range 0 to 2π we need to find values of $x - \frac{\pi}{3}$

in the range $-\frac{\pi}{3}$ to $2\pi - \frac{\pi}{3}$,

$\cos\left(x - \frac{\pi}{3}\right) = \frac{1}{2} \Rightarrow x - \frac{\pi}{3} = -\frac{\pi}{3}, \frac{\pi}{3}, \frac{5\pi}{3}$

$\therefore \quad x = 0, \frac{2\pi}{3}, 2\pi$

Exercise 2.7b

1 Express $3\cos x - 4\sin x$ in the form $r\cos(x + \alpha)$.

Hence find the maximum and minimum values of $3\cos x - 4\sin x$ and the values of x at which they occur in the range $0 \le x \le 2\pi$

2 Solve the equation $\cos x + \sin x = \sqrt{2}$ for $0 \le x \le 2\pi$

Learning outcomes

- To prove trigonometric identities
- To solve trigonometric equations in a given range

You need to know

- The properties of the trigonometric ratios

The important identities

In previous pages we have introduced some trigonometric identities one group at a time.

We now collect them together:

$$\tan \theta \equiv \frac{\sin \theta}{\cos \theta}$$

$$\cos^2 \theta + \sin^2 \theta \equiv 1$$

$$1 + \tan^2 \theta \equiv \sec^2 \theta$$

$$\cot^2 \theta + 1 \equiv \operatorname{cosec}^2 \theta$$

$$\sin (A + B) \equiv \sin A \cos B + \cos A \sin B$$

$$\sin (A - B) \equiv \sin A \cos B - \cos A \sin B$$

$$\cos (A + B) \equiv \cos A \cos B - \sin A \sin B$$

$$\cos (A - B) \equiv \cos A \cos B + \sin A \sin B$$

$$\tan(A + B) \equiv \frac{\tan A + \tan B}{1 - \tan A \tan B}$$

$$\tan(A - B) \equiv \frac{\tan A - \tan B}{1 + \tan A \tan B}$$

$$\sin 2A \equiv 2 \sin A \cos A$$

$$\tan 2A \equiv \frac{2 \tan A}{1 - \tan^2 A}$$

$$\cos 2A \equiv \begin{Bmatrix} \cos^2 A - \sin^2 A \\ 2 \cos^2 A - 1 \\ 1 - 2 \sin^2 A \end{Bmatrix}$$

$$\cos^2 A = \tfrac{1}{2}(1 + \cos 2A)$$

$$\sin^2 A = \tfrac{1}{2}(1 - \cos 2A)$$

$$2 \sin A \cos B \equiv \sin (A + B) + \sin (A - B)$$

$$2 \cos A \sin B \equiv \sin (A + B) - \sin (A - B)$$

$$2 \cos A \cos B \equiv \cos (A + B) + \cos (A - B)$$

$$-2 \sin A \sin B \equiv \cos (A + B) - \cos (A - B)$$

$$\sin P + \sin Q \equiv 2 \sin \tfrac{1}{2}(P + Q) \cos \tfrac{1}{2}(P - Q)$$

$$\sin P - \sin Q \equiv 2 \cos \tfrac{1}{2}(P + Q) \sin \tfrac{1}{2}(P - Q)$$

$$\cos P + \cos Q \equiv 2 \cos \tfrac{1}{2}(P + Q) \cos \tfrac{1}{2}(P - Q)$$

$$\cos P - \cos Q \equiv -2 \sin \tfrac{1}{2}(P + Q) \sin \tfrac{1}{2}(P - Q)$$

Proving trigonometric identities

To prove an identity, it is sensible to work on one side at a time.

The examples and exercises in the previous pages have all been associated with a specific set of identities. In Exercise 2.8a you are asked to prove mixed identities. Some guidelines on where to start are:

- work with the more complicated side

- convert all ratios to sine and cosine

- when a multiple angle is involved, start with that and break it down to ratios of a single angle using the compound angle formulae for odd multiples (e.g. $\sin 3A$), and the double angle formulae for even multiples (e.g. $\sin 2A$)

- use the factor formulae to express a product of ratios as a sum, or vice versa.

The following example illustrates that it may be necessary to work on both sides, but separately.

 Exam tip

Remember that these points are only guidelines, and practice will help you develop strategies that work for you.

Example

Prove that $\sin 3A = (\sin 2A - \sin A)(1 + 2\cos A)$

There are multiple angles on both sides, so we will start with the RHS and express that in terms of ratios of the single angle A and simplify.

RHS: $(\sin 2A - \sin A)(1 + 2\cos A) = (2\sin A\cos A - \sin A)(1 + 2\cos A)$

$$= \sin A(2\cos A - 1)(1 + 2\cos A)$$

$$= \sin A(4\cos^2 A - 1)$$

We now turn to the LHS and express that in terms of ratios of the single angle A.

LHS: $\sin 3A = \sin 2A\cos A + \cos 2A\sin A$

$$= 2\sin A\cos^2 A + \sin A(2\cos^2 A - 1)$$

This form for $\cos 2A$ is chosen because we have the factor $\sin A$ and we want the other factor to involve $\cos A$

$$= \sin A(2\cos^2 A + 2\cos^2 A - 1)$$

$$= \sin A(4\cos^2 A - 1) = \text{RHS}$$

Therefore $\sin 3A = (\sin 2A - \sin A)(1 + 2\cos A)$

Exercise 2.8a

Prove the following identities.

1 $(\cot A + \operatorname{cosec} A)^2 \equiv \dfrac{1 + \cos A}{1 - \cos A}$

2 $\cot 2A + \operatorname{cosec} 2A \equiv \cot A$

3 $\dfrac{\cos A}{1 - \tan A} + \dfrac{\sin A}{1 - \cot A} \equiv \sin A + \cos A$

4 $\sin(\theta + \frac{\pi}{3}) \equiv \sin(\frac{2\pi}{3} - \theta)$

5 $\dfrac{1 - \cos 2A}{\sin 2A} \equiv \tan A$

6 $\dfrac{\cos 2A + \cos 2B}{\cos 2B - \cos 2A} \equiv \cot(A + B)\cot(A - B)$

7 $\cos 4x \equiv 8\cos^4 x - 8\cos^2 x + 1$

8 $\sec^2 \theta \equiv \dfrac{\operatorname{cosec} \theta}{\operatorname{cosec} \theta - \sin \theta}$

9 $\cot(A + B) \equiv \dfrac{\cot A\cot B - 1}{\cot A + \cot B}$

10 $\dfrac{\sin 3\theta + \sin 5\theta}{\sin 4\theta + \sin 6\theta} \equiv \dfrac{\sin 4\theta}{\sin 5\theta}$

11 $\cos 3A \equiv 4\cos^3 A - 3\cos A$

12 $\tan(A + B) - \tan A \equiv \dfrac{\sin B}{\cos A\cos(A + B)}$

13 $\dfrac{\cos x - \cos y}{\sin x + \sin y} \equiv \tan\frac{1}{2}(x + y)$

Solving trigonometric equations in a given range

Any of the trig identities, together with the transformation $a \cos x + b \sin x = r \sin/\cos(x + \alpha)$, can be used to help solve a trig equation.

There are two general approaches to solving trig equations. First see whether the equation will factorise, either in its given form or by using an identity. If this is not possible (and it often is not), try to reduce the equation to a form that involves only one trig ratio of one angle.

The tables that follow list most of the common categories of trig equations, together with an appropriate method for solving them. The lists are not exhaustive, nor are they infallible.

Equations containing one angle only

Equation category	Method
1 $a \cos x + b \sin x = 0$	Divide by $\cos x$, provided that $\cos x \neq 0$
2 $a \cos x + b \sin x = c$	Transform the LHS to $r \cos(x - \alpha)$
3 $a \cos^2 x + b \sin x = c$ $a \sin^2 x + b \cos x = c$ $a \tan^2 x + b \sec x = c$	Use the Pythagorean identities to express the LHS in terms of one ratio only
4 $a \cos x + b \tan x = 0$ $a \sin x + b \tan x = 0$	Multiply by $\cos x$, provided that $\cos x \neq 0$

Equations containing multiples of one angle

Equation category	Method
1 $a \cos x + b \cos 2x = c$ $a \sin x + b \cos 2x = c$	Use the double angle formulae to reduce to an equation containing trig ratios of only x
2 $\cos ax = c$ $\sin ax = c$	Solve for ax in a times the required range and then divide by a

The methods listed do not give the only way of solving a particular equation, nor do they always lead to the quickest solution. An equation may be able to be simplified quickly when part of it is recognised as part of a trig identity. Sometimes you may need to classify each side of an equation independently.

Practice will help you recognise the best way to tackle any equation.

Example

Solve the equation $\sin 2\theta \cos \theta + \cos 2\theta \sin \theta = \tan 3\theta$ for values of θ in the range $0 \leqslant \theta \leqslant \dfrac{\pi}{2}$

$\sin 2\theta \cos \theta + \cos 2\theta \sin \theta = \sin 3\theta$ The LHS is the expansion of $\sin(2\theta + \theta)$

$\Rightarrow \;\; \sin 3\theta = \tan 3\theta$

$\Rightarrow \;\; \sin 3\theta = \dfrac{\sin 3\theta}{\cos 3\theta}$

$\Rightarrow \;\; \sin 3\theta - \sin 3\theta \sec 3\theta = 0$

$\Rightarrow \;\; \sin 3\theta \left(1 - \sec 3\theta\right) = 0$

$\Rightarrow \;\; \sin 3\theta = 0$ or $\sec 3\theta = 1$

For values of θ in the range $0 \leqslant \theta \leqslant \dfrac{\pi}{2}$, we need to find values of 3θ in the range $0 \leqslant \theta \leqslant \dfrac{3\pi}{2}$,

$\quad \sin 3\theta = 0 \;\Rightarrow\; 3\theta = 0$ or π and $\sec 3\theta = 1 \;\Rightarrow\; 3\theta = 0$

$\therefore \quad \theta = 0, \dfrac{\pi}{3}$

Example

Solve the equation $\cos 4\theta + \cos 2\theta + \cos 3\theta = 0$ for $0 \leqslant \theta \leqslant \pi$

$\quad \cos 4\theta + \cos 2\theta + \cos 3\theta = 0$

Using the factor formula on the first two terms gives a factor $\cos 3\theta$.

This gives $\cos 3\theta$ as a factor of the LHS.

$\Rightarrow \;\; 2\cos 3\theta \cos \theta + \cos 3\theta = 0$

$\Rightarrow \;\; \cos 3\theta (2\cos \theta + 1) = 0$

$\Rightarrow \;\; \cos 3\theta = 0$ or $\cos \theta = -\dfrac{1}{2}$

$\therefore \quad 3\theta = \dfrac{\pi}{2}, \dfrac{3\pi}{2}, \dfrac{5\pi}{2}$ so $\theta = \dfrac{\pi}{6}, \dfrac{\pi}{2}, \dfrac{5\pi}{6}$

$\quad$ or $\theta = \dfrac{2\pi}{3}$

i.e. $\theta = \dfrac{\pi}{6}, \dfrac{\pi}{2}, \dfrac{2\pi}{3}, \dfrac{5\pi}{6}$

Exercise 2.8b

Give answers that are not exact correct to three significant figures.

Solve the following equations for values of θ in the range $0 \leqslant \theta \leqslant \pi$

1 $\cos 2\theta \sin \theta + \sin 2\theta \cos \theta = 1$

2 $\sin^2 \theta - 2\cos \theta = 1$

3 $5\sin \theta + 12\cos \theta = 13$

4 $\sin 3\theta - \sin \theta = 0$

5 $\cos^2 \theta + 2\sin^2 \theta = 2$

6 $\sin 5\theta - \sin \theta + \cos 3\theta = 0$

Solve the following equations for values of θ in the range $0 \leqslant \theta \leqslant 2\pi$

7 $2\sec^2 \theta + \tan \theta = 3$

8 $4\sin^2 \theta - 2\cos^2 \theta = 4\cos^2 \theta - 1$

9 $\dfrac{1}{1 + \cos \theta} + \dfrac{1}{1 - \cos \theta} = 4$

10 $\tan 2\theta - \cot 2\theta = 0$

11 $\sin 3\theta + \sin \theta = \cos \theta + \cos 3\theta$

12 $4\sin \theta = \sec \theta$

Learning outcomes

- To find the general solution of trig equations

You need to know

- The properties of the trigonometric functions
- The main trigonometric identities

General solutions of trigonometric equations

The equation $\cos x = 1$ has a finite number of solutions in a given range of values of x.

All possible values of $\cos x$ occur in the range $-\pi \leqslant x \leqslant \pi$ and $\cos x = 1$ has one solution in this range, namely $x = 0$

However, $\cos x = 1$ for an infinite number of values of x.
In fact $\cos x = 1$ for every multiple (both positive and negative) of 2π.

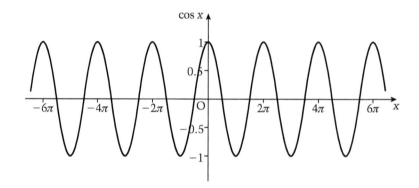

Therefore when the range of values of x is unrestricted, the solutions of $\cos x = 1$ can be written as
$x = 2n\pi$ for $n \in \mathbb{Z}$.

This is called the **general solution** of the equation.

The general solution of $\cos x = c$, $|c| \leqslant 1$

Within the range $-\pi \leqslant x \leqslant \pi$, the equation $\cos x = c$ has, in general, two solutions.

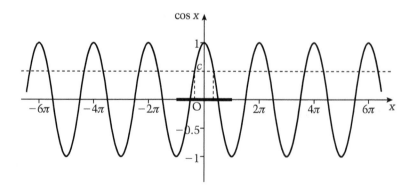

If one solution is $x = \alpha$, then the other solution is $x = -\alpha$

The graph of $f(x) = \cos x$ has a pattern that repeats every interval of 2π. Therefore we can get the general solution of the equation by adding multiples of 2π to α and to $-\alpha$.

We can write this solution as $x = 2n\pi \pm \alpha$, where n is an integer.

Therefore the general solution of the equation $\cos x = c$ is
$$x = 2n\pi \pm \alpha \text{ for } n \in \mathbf{Z}$$
where α is a solution in the range $-\pi \leqslant x \leqslant \pi$

Example

Find the general solution of the equation $\cos x = \frac{1}{2}$

$\cos x = \frac{1}{2}$

$\Rightarrow x = \pm\frac{\pi}{3}$ in the range $-\pi \leqslant x \leqslant \pi$

Therefore the general solution is

$x = 2n\pi \pm \frac{\pi}{3}$

The general solution of $\sin x = c$, $|c| \leqslant 1$

All possible values of $\sin x$ occur in the range $0 \leqslant x \leqslant 2\pi$ and, in general, the equation $\sin x = c$ has two solutions in this range.

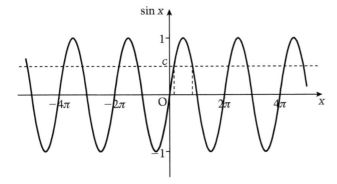

If the smaller solution is $x = \alpha$, then the other solution is $\pi - \alpha$

The graph of $f(x) = \sin x$ has a pattern that repeats every interval of 2π.

Therefore we can get the general solution of $\sin x = c$ by adding multiples of 2π to α and to $\pi - \alpha$

We can write this solution as $x = 2n\pi + \alpha$ and $x = 2n\pi + \pi - \alpha$ where n is an integer.

Therefore the general solution of the equation $\sin x = c$ is
$$x = 2n\pi + \alpha \text{ and } x = (2n + 1)\pi - \alpha \text{ for } n \in \mathbf{Z}$$
where α is the smallest solution in the range $0 \leqslant x \leqslant 2\pi$

Example

Find the general solution of the equation $\sin x = \frac{\sqrt{2}}{2}$

$\sin x = \frac{\sqrt{2}}{2}$

$\Rightarrow x = \frac{\pi}{4}$ as the smallest solution for values of x in the range

$0 \leqslant x \leqslant 2\pi$

Therefore the general solution is $x = 2n\pi + \frac{\pi}{4}$ or $x = (2n + 1)\pi - \frac{\pi}{4}$

The general solution of $\tan x = c$

All possible values of $\tan x$ occur in the range $-\frac{\pi}{2} \leqslant x \leqslant \frac{\pi}{2}$ and the equation $\tan x = c$ has one solution in this range.

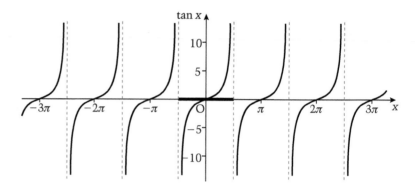

The graph of $f(x) = \tan x$ has a pattern that repeats every interval of π, so if $x = \alpha$ is the solution of $\tan x = c$ in the range $-\frac{\pi}{2} \leqslant x \leqslant \frac{\pi}{2}$, the general solution can be written as

$$x = n\pi + \alpha, \, n \in \mathbb{Z}.$$

Therefore the general solution of the equation $\tan x = c$ is
$$x = n\pi + \alpha, \, n \in \mathbf{Z}$$

where α is the solution in the range $-\frac{\pi}{2} \leqslant x \leqslant \frac{\pi}{2}$

Example

Find the general solution of the equation $\tan x = -1$

$\tan x = -1 \Rightarrow x = -\frac{\pi}{4}$ for values of x in the range $-\frac{\pi}{2} \leqslant x \leqslant \frac{\pi}{2}$

Therefore the general solution is $x = n\pi - \frac{\pi}{4}$

When a multiple angle is involved, first find the general solution for the multiple angle. Then use that to find the general solution for the single angle.

Example

Find the general solution of the equation $\sin 2x = \frac{1}{2}$

When $\sin 2x = \frac{1}{2}$, the solution for $2x$ in the range $0 \leqslant 2x \leqslant 2\pi$ is

$2x = \frac{\pi}{6}$ and $2x = \frac{5\pi}{6}$

Therefore the general solution for $2x$ is

$2x = \frac{\pi}{6} + 2n\pi$ and

$2x = \frac{5\pi}{6} + 2n\pi$

giving the general solution for x as

$x = \frac{\pi}{12} + n\pi$ and $x = \frac{5\pi}{12} + n\pi$

Example

Find the general solution of the equation

$$\sin x + 2\sin x \cos 2x = 0$$

$$\sin x + 2\sin x \cos 2x = 0$$
$$\Rightarrow \sin x\,(1 + 2\cos 2x) = 0$$

$\therefore$ Either $\sin x = 0 \Rightarrow x = 0,\ \pi,$ or 2π in the interval $0 \leqslant x \leqslant 2\pi$

Then the general solution is $x = n\pi$

Or $\cos 2x = -\dfrac{1}{2} \Rightarrow 2x = \pm\dfrac{2\pi}{3}$ in the interval $-\pi \leqslant x \leqslant \pi$

Then the general solution is $2x = 2n\pi \pm \dfrac{2\pi}{3} \Rightarrow x = n\pi \pm \dfrac{\pi}{3}$

i.e. $x = n\pi$ or $x = n\pi \pm \dfrac{\pi}{3}$

Example

Find the general solution of the equation

$$\cos\frac{x}{2} - \sin\frac{x}{2} = 1$$

$\cos\dfrac{x}{2} - \sin\dfrac{x}{2} = r\cos\left(\dfrac{x}{2} + \alpha\right)$ where $r = \sqrt{2}$ and $\tan\alpha = 1$

$\therefore\ \cos\dfrac{x}{2} - \sin\dfrac{x}{2} = 1$

$\Rightarrow \cos\left(\dfrac{x}{2} + \dfrac{\pi}{4}\right) = \dfrac{1}{\sqrt{2}}$

For values of x in the range $-\pi \leqslant x \leqslant \pi$,

we need values of $\left(\dfrac{x}{2} + \dfrac{\pi}{4}\right)$ from $-\dfrac{\pi}{2} + \dfrac{\pi}{4}$ to $\dfrac{\pi}{2} + \dfrac{\pi}{4}$

$\left(\dfrac{x}{2} + \dfrac{\pi}{4}\right) = -\dfrac{\pi}{4}$ or $\dfrac{\pi}{4}$

$\Rightarrow x = -\pi$ or 0

Therefore the general solution is is $x = n\pi$

Exercise 2.9

Find the general solutions of the equations.

1 $\sin x = 1$

2 $\cos x = -1$

3 $\tan x = 1$

4 $\sqrt{2}\cos x = 1$

5 $\cos 3x = \frac{1}{2}$

6 $\sin^2 x + \sin x = 0$

7 $\tan 2x = \sqrt{3}$

8 $\sin 3x + \sin x = 0$

Learning outcomes

- To revise the gradient of a straight line
- To find the angle between two straight lines
- To find the distance of a point from a straight line

You need to know

- The Pythagorean identities
- The compound angle formulae
- The exterior angle property of a triangle

The gradient of a straight line

The gradient of a straight line is given by finding the increase in y divided by the increase in x when moving from one point on the line to another point on the line.

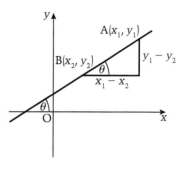

In the diagram, the gradient of the line is $\dfrac{y_1 - y_2}{x_1 - x_2}$. This is also the value of $\tan \theta$.

Therefore the gradient of a straight line is equal to the tangent of the angle that the line makes with the positive direction of the x-axis.

When two lines are perpendicular, the product of their gradients is -1. Therefore if the equation of a line L is $y = mx + c$, any line perpendicular to L will have an equation of the form $y = -\dfrac{1}{m}x + k$

The angle between two lines

The lines L_1 and L_2 have gradients m_1 and m_2 respectively, where $m_1 = \tan \theta_1$ and $m_2 = \tan \theta_2$

The angle α between the lines is $\theta_1 - \theta_2$.

Therefore $\tan \alpha = \tan(\theta_1 - \theta_2)$

$$= \frac{\tan \theta_1 - \tan \theta_2}{1 + \tan \theta_1 \tan \theta_2}$$

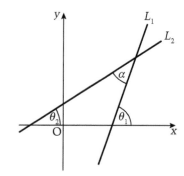

Therefore the angle, α, between lines with gradients m_1 and m_2 is given by $\tan \alpha = \dfrac{m_1 - m_2}{1 + m_1 m_2}$

Example

L_1 is the line $y = 3x - 5$ and L_2 is the line $2x + y + 1 = 0$. What is the angle between the lines?

$m_1 = 3$ and $m_2 = -2$, so the angle between the lines is given by

$$\tan \alpha = \frac{3 - (-2)}{1 + (3)(-2)} = -\frac{5}{5} \Rightarrow \alpha = \frac{3\pi}{4}$$

This is the obtuse angle between the lines; the acute angle is

$$\pi - \frac{3\pi}{4} = \frac{\pi}{4}$$

The distance of a point from a line

The distance of a point from a line is the perpendicular distance.

In the diagram, the distance of the point A(a, b) from the line $y = mx + c$ is the length of the line AN.

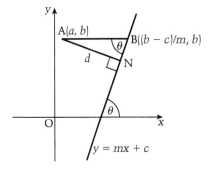

B is a point on the line such that AB is parallel to the x-axis, so b is the y-coordinate of B.

Therefore from the equation of the line the x-coordinate of B is $\dfrac{b-c}{m}$

This gives the length of AB as $\left|\dfrac{b-c}{m} - a\right| = \left|\dfrac{b-c-ma}{m}\right|$

$\qquad$ AN $= d =$ AB $\sin\theta$

But $\tan\theta = m$, therefore $\sin\theta = \dfrac{m}{\sqrt{m^2 + 1}}$ $\quad$ Using $1 + \cot^2\theta = \operatorname{cosec}^2\theta$

Hence $d = \left|\dfrac{b-c-ma}{\sqrt{m^2 + 1}}\right|$

Note that in the diagram point A is on the left of the line. If A were on the right of the line, then we would write $a - (b-c)/m$ for the length of AB. The modulus is needed because the length of AB is the *difference* between the x-coordinates of A and B.

Note also that if θ is obtuse, the angle ABN $= \pi - \theta$
But $\sin(\pi - \theta) = \sin\theta$, so the result is valid when the line has a negative gradient.

The distance of the point (a, b) from the line $y = mx + c$

is given by $\left|\dfrac{b - c - ma}{\sqrt{m^2 + 1}}\right|$

Example

Find the distance of the point $(2, -3)$ from the line $2x - 3y + 4 = 0$

Rearranging the equation of the line as $y = \frac{2}{3}x + \frac{4}{3}$

$\Rightarrow a = 2, b = -3, m = \frac{2}{3}$ and $c = \frac{4}{3}$

Therefore the distance of $(2, -3)$ from $2x - 3y + 4 = 0$ is given by

$\left|\dfrac{-3 - \frac{4}{3} - \frac{4}{3}}{\sqrt{(\frac{2}{3})^2 + 1}}\right| = \dfrac{17\sqrt{13}}{13}$

Exercise 2.10

1 Find the acute angle between the lines whose equations are
$y = 3x - 2$ and $2x + 3y - 1 = 0$

2 A is the point $(2, 5)$ and B is the point $(-5, 2)$. Find the equation of the line that bisects the angle AOB, where O is the origin.

3 Find the distance of the point $(1, 5)$ from the line whose equation is
$y = 3 - x$

4 A point P(a, b) is equidistant from the line $2y = 5x - 1$ and from the point $(1, 1)$.

Find a relationship between a and b.

Learning outcomes

- To define the meaning of loci in the context of the *xy*-plane
- To find the Cartesian equation of a given locus
- To find the Cartesian equation of a circle

You need to know

- How to find the distance between two points
- How to find the equation of a straight line
- How to form a perfect square

Loci

When a condition is placed on the possible positions of a point P, then P is restricted to a particular set of points. This set of points is called the *locus* of P.

When P is the point (x, y), the relationship between x and y that satisfies this condition is called the *Cartesian equation* of the locus of P.

For example, if A is the point $(1, 4)$ and the condition on P(x, y) is that the gradient of AP is 2, then the gradient of AP is $\dfrac{y - 4}{x - 1}$

∴ the Cartesian equation of the locus of

P is $\qquad \dfrac{y - 4}{x - 1} = 2$

$\qquad\qquad \Rightarrow y = 2x + 2$

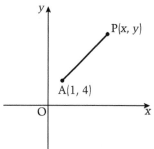

Example

A point P(x, y) is the same distance from the point $(1, 2)$ as it is from the line $x = 3$. Find the equation of the locus of P.

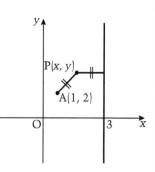

The distance of P from the line $x = 3$ is $3 - x$

The distance of P from the point $(1, 2)$ is $\sqrt{(x - 1)^2 + (y - 2)^2}$

∴ the equation of the locus of P is given by $3 - x = \sqrt{(x - 1)^2 + (y - 2)^2}$

$\Rightarrow (3 - x)^2 = (x - 1)^2 + (y - 2)^2$

$\Rightarrow 9 - 6x + x^2 = x^2 - 2x + 1 + y^2 - 4y + 4$

$\Rightarrow y^2 - 4y + 4x - 4 = 0$

Exercise 2.11a

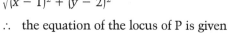

Find the Cartesian equation of the locus of P(x, y) when P satisfies the following conditions.

1 P is the same distance from the point $(0, 4)$ and the line $x = 6$

2 P is the same distance from the points $(1, 2)$ and $(-2, 4)$.

3 P is twice the distance from the line $y = 5$ as it is from the point $(2, 0)$.

The Cartesian equation of a circle

A circle is the locus of points that are a constant distance from a fixed point.

If $P(x, y)$ is any point on the circle of radius r and centre $C(p, q)$, then

$CP = r$ and $CP = \sqrt{(x - p)^2 + (y - q)^2}$

$\therefore \quad (x - p)^2 + (y - q)^2 = r^2$

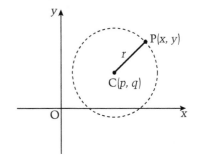

> **i.e. the Cartesian equation of a circle, centre (p, q) and
> radius r is $(x - p)^2 + (y - q)^2 = r^2$**

Therefore the equation of a circle, centre $(2, -1)$ and radius 3, is given by

$(x - 2)^2 + (y - (-1))^2 = 9 \Rightarrow x^2 + y^2 - 4x + 2y - 4 = 0$

Conversely the circle whose equation is $(x + 1)^2 + (y - 3)^2 = 8$ has its centre at $(-1, 3)$ and its radius is $\sqrt{8}$.

An equation of the form $\quad x^2 + y^2 + 2gx + 2fy + c = 0 \qquad$ [1]

where f, g and c are constants, can be rearranged as

$$x^2 + 2gx + g^2 + y^2 + 2fy + f^2 + c = g^2 + f^2$$

$\Rightarrow \qquad\qquad\qquad (x + g)^2 + (y + f)^2 = g^2 + f^2 - c \quad$ [2]

> **Comparing [1] and [2] shows that
> $x^2 + y^2 + 2gx + 2fy + c = 0$ is the equation of a circle
> with centre $(-g, -f)$ and radius $\sqrt{g^2 + f^2 - c}$
> provided that $\sqrt{g^2 + f^2 - c}$ is a real number.**

Notice that the coefficients of x^2 and y^2 are equal and that there is no xy term.

Example

Find the centre and radius of the circle whose equation is
$x^2 + y^2 + 2x - 4y + 4 = 0$

Rearranging $x^2 + y^2 + 2x - 4y + 4 = 0$ as
$(x + 1)^2 + (y - 2)^2 = 1 + 4 - 4 = 1$ gives the centre as the point
$(-1, 2)$ and the radius as 1.

Alternatively comparing $x^2 + y^2 + 2x - 4y + 4 = 0$ with
$x^2 + y^2 + 2gx + 2fy + c = 0$ gives $g = 1, f = -2$ and $c = 4$

Then using $(-g, -f)$ and $\sqrt{g^2 + f^2 - c}$, gives the centre as the point
$(-1, 2)$ and the radius as $\sqrt{1 + 4 - 4} = 1$

Exercise 2.11b

1. Find the equation of the circle whose centre is $(3, -2)$ and whose radius is 5.

2. Find the centre and radius of the circle whose equation is
   $x^2 + y^2 - 6x + 2y - 6 = 0$

3. Find the centre and radius of the circle whose equation is
   $2x^2 + 2y^2 + 6x + 4y + 1 = 0$

   (Hint: divide the equation by 2.)

4. Explain why the equation $x^2 + y^2 + 2x - y + 6 = 0$ cannot be the equation of a circle.

Learning outcomes

- To find the equations of tangents and normals to circles
- To find the condition for a line to be a tangent to a circle

You need to know

- How to find the centre and radius from the Cartesian equation of a circle
- The basic facts about the geometry of a circle
- How to find the points of intersection of a line and a curve
- How to find the distance of a point from a line

The equation of a tangent to a circle at a given point

When we know the coordinates of the centre and the radius of a circle, we can use the fact that the tangent at a given point on the circle is perpendicular to the radius through the point of contact.

> ### Example
> Find the equation of the tangent at the point A$(3, -11)$ on the circle
> $$x^2 + y^2 - 4x + 10y - 8 = 0$$
> First rearrange the equation as
> $$(x - 2)^2 + (y + 5)^2 = 8 + 4 + 25$$
> Therefore the centre of the circle is the point C$(2, -5)$
> The gradient of AC $= -6$
> $\therefore$ the gradient of the tangent at A is $\frac{1}{6}$
> So the equation of the tangent at A is $y - (-11) = \frac{1}{6}(x - 3)$
> $\Rightarrow \qquad\qquad\qquad\qquad\qquad x - 6y - 69 = 0$

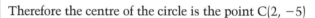

The normal to a curve

The normal to a curve is the line perpendicular to the tangent to a curve through the point of contact.

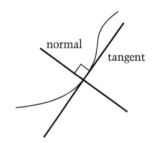

In the case of a circle, the normal is the line containing the radius through the point of contact.

The condition for a line to be a tangent to a circle

There are two methods for determining whether a line is a tangent to a circle.

The first method uses the fact that a line is a tangent to a circle if the distance from the centre of the circle to the line is equal to the radius of the circle.

The second method uses the fact that there will be a repeated root when the equations of the line and the circle are solved simultaneously.

Example

(a) Find the values of c for which the line $x - y + c = 0$ is a tangent to the circle $x^2 + y^2 + 2x - 8 = 0$

(b) Find the equation of the diameter of the circle that is parallel to the tangent.

(a) $x^2 + y^2 + 2x - 8 = 0 \Rightarrow (x + 1)^2 + y^2 = 9$

∴ the point $(-1, 0)$ is the centre of the circle and the radius is 3.

First method
For the line $x - y + c = 0$ (i.e. $y = x + c$) to be a tangent to the circle, the distance of the line from $(-1, 0)$ is 3.

∴ using $d = \left| \dfrac{b - c - ma}{\sqrt{m^2 + 1}} \right|$, where $d = 3$, $a = -1$, $b = 0$ and

$m = 1$ gives

$$3 = \left| \frac{0 - c - (1)(-1)}{\sqrt{1 + 1}} \right| \Rightarrow 3 = \pm \frac{1 - c}{\sqrt{2}} \Rightarrow c = 1 \pm 3\sqrt{2}$$

Second method
Solving $x - y + c = 0$ and $(x + 1)^2 + y^2 = 9$ simultaneously gives

$$(x + 1)^2 + (x + c)^2 = 9$$

$$\Rightarrow \qquad 2x^2 + 2x(1 + c) + c^2 - 8 = 0$$

For the line to be a tangent this equation must have equal roots,

i.e. '$b^2 - 4ac = 0$' $\Rightarrow 4(1 + c)^2 - 8c^2 + 64 = 0$

$$\Rightarrow \qquad\qquad c^2 - 2c - 17 = 0$$

$$\Rightarrow \qquad\qquad c = \frac{2 \pm \sqrt{72}}{2} = 1 \pm 3\sqrt{2}$$

(b) The diameter goes through the centre of the circle, i.e. through $(-1, 0)$.

The diameter is parallel to the tangents so its gradient is 1.

Therefore the equation of the diameter is $(y - 0) = 1(x - (-1))$, i.e. $y = x + 1$

Exercise 2.12

1 Find the equations of the tangents to the circle $x^2 + y^2 + 6y - 11 = 0$ at the points on the circle where $y = 1$

2 Determine whether the line $3x - 4y + 5 = 0$ is a tangent to the circle $x^2 + y^2 + 4x + 8y = 0$

3 **(a)** Explain why the circles $(x + 1)^2 + (y + 2)^2 = 9$ and $x^2 + y^2 - 4x + 12y + 36 = 0$ touch.

(b) Find the coordinates of the point of contact of the two circles.

(c) Find the equation of the common normal to the circles through the point of contact.

4 Find the condition that m and c satisfy if the line $y = mx + c$ is a tangent to the circle whose equation is $x^2 + y^2 + 6x + 5 = 0$

Learning outcomes

- To define a parameter
- To find the Cartesian equation of a curve given in parametric form

You need to know

- How to find the centre and radius of a circle from its equation
- The Pythagorean trig identities and the double angle identities

The definition of a parameter

When a direct relationship between x and y is difficult to work with, it is often easier to express each of x and y in terms of a third variable. This variable is called a *parameter*.

For example, in the equations $x = t^2$, $y = t - 1$, t is the parameter. The equations are called the *parametric equations* of the curve.

A point P(x, y) is on the curve given by these equations if and only if the coordinates of P are $(t^2, t - 1)$.

By giving t any value we choose, we get a pair of corresponding values of x and y. For example, when $t = 2$, $x = 4$ and $y = 1$. Therefore $(4, 1)$ is a point on this curve.

By giving t several other values we can plot points and draw the curve.

t	-3	-2	-1	0	1	2	3
x	9	4	1	0	1	4	9
y	-4	-3	-2	-1	0	1	2

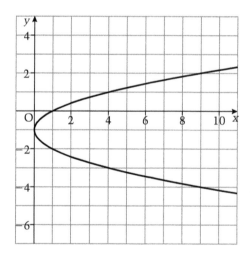

The relationship between parametric equations and Cartesian equations

The Cartesian equation of a curve can be found by eliminating the parameter.

In the case of the equations $x = t^2$, $y = t - 1$, eliminating t gives $x = (y + 1)^2$

When the parametric equations involve trigonometric ratios of an angle θ, the trig identities are useful to help eliminate θ.

For example, a curve has these parametric equations:

$x = 2 \sin \theta$ [1]   and $y = 3 \cos 2\theta$ [2]

The identity $\cos 2\theta \equiv 1 - 2\sin^2 \theta$ can be used to find the Cartesian equation of the curve.

$[1] \Rightarrow \sin \theta = \frac{x}{2}$ and $[2] \Rightarrow \cos 2\theta = \frac{y}{3}$

Therefore $\frac{y}{3} = 1 - 2\left(\frac{x}{2}\right)^2$

$\Rightarrow \qquad 2y = 6 - 3x^2$

It is not always easy to convert a Cartesian equation to parametric equations. However, the Cartesian equation of a circle can be converted to parametric equations.

For example, the circle whose equation is $(x - 2)^2 + (y - 1)^2 = 9$ has radius 3 and centre $(2, 1)$.

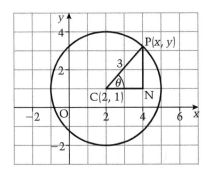

From the diagram, $CN = 3\cos \theta$

therefore $\qquad x = 2 + 3\cos \theta$

$\qquad\qquad PN = 3\sin \theta$

therefore $\qquad y = 1 + 3\sin \theta$

Hence, $x = 2 + 3\cos \theta$ and $y = 1 + 3\sin \theta$ are the parametric equations of the circle, where θ is the parameter.

Exercise 2.13

1   Find the Cartesian equation of the following curves whose equations are given parametrically.

   (a) $x = t^2 + 1, y = t + 2$

   (b) $x = \dfrac{1}{1 + t}, y = \dfrac{t - 1}{t^2}$

   (c) $x = \sec \theta, y = 3\tan \theta$

2   Show that the curve whose parametric equations are

$$x = 3(1 + \sin \theta) \quad \text{and} \quad y = 3\cos \theta$$

   represents a circle.
   Give the coordinates of the centre and the radius of the circle.

Conic sections

There is a set of curves called conic sections that come from the intersection of a plane and a right circular cone.

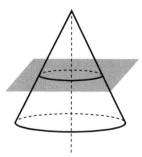

When the plane is perpendicular to the axis of the cone, the curve is a *circle*. (The circle is not always considered to be one of the conic sections.)

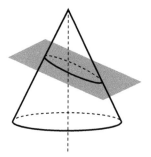

When the plane is at an angle less than the inclination of the slant of the curved surface, the curve is an *ellipse*.

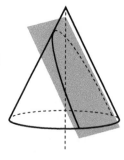

When the plane is parallel to the slant of the curved surface, the curve is a *parabola*.

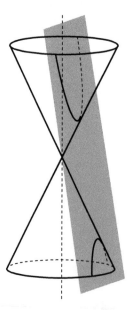

When the cone is double-ended and the plane is at an angle greater than the slant of the curved surface, a two-part curve is formed called a *hyperbola*.

Like many mathematical activities, conic sections were investigated as a purely academic activity with no interest in their applications. And like many such mathematical activities, they end up having wide scientific uses.

It was in the 17th century that Kepler discovered that the orbits of the planets round the Sun are ellipses. In fact, our moon and satellites move round the Earth in elliptical orbits.

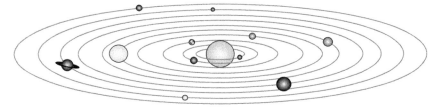

Around the same time Galileo found that an object projected at an angle to the vertical follows a path whose shape is a parabola. For example, the flight of a cricket ball thrown to another player is a parabola.

One of the most important scientific applications is the use of parabolic mirrors in giant telescopes to focus light. (A parabolic surface is made by rotating a parabola about its axis.)

This comes from a property of the parabola that means all light coming in parallel to the axis is reflected onto one point (called the focus).

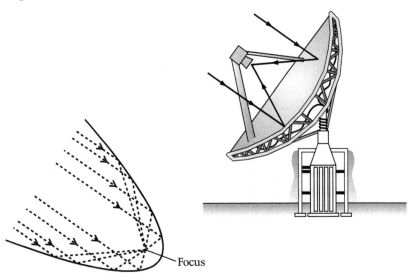

Focus

The Hubble space telescope currently orbiting the Earth has a parabolic mirror which collects starlight.

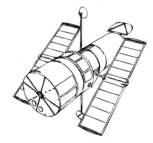

In the next topics we use coordinate geometry to find equations for the parabola and ellipse.

- To define a parabola as a locus
- To find the Cartesian equation and the parametric equations of the parabola

You need to know

- The meaning of a parameter
- How to find the Cartesian equation of a locus
- How to find the distance between two points
- The condition for a line to be a tangent to a curve

The parabola

We are already familiar with the shape of a parabola – the graph of $y = ax^2 + bx + c$ is a parabola.

One of the properties of a parabola discovered by the ancient Greeks is that any point on a parabola is equidistant from a fixed point (called the *focus*) and a fixed straight line (called the *directrix*).

We can now use this property to derive the Cartesian equation of the standard parabola.

The simplest equation is obtained by taking the fixed point as A$(a, 0)$ and the fixed line as $x = -a$, then P(x, y) is such that PA = PN

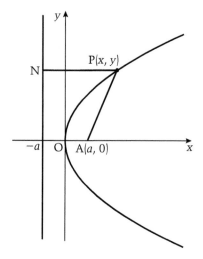

$$PA^2 = (x - a)^2 + y^2 \text{ and } PN = (x + a) \Rightarrow PN^2 = (x + a)^2$$
$$\therefore \qquad (x - a)^2 + y^2 = (x + a)^2$$
$$\Rightarrow \qquad y^2 = 4ax$$

$y^2 = 4ax$ is the Cartesian equation of the standard parabola whose vertex is at the origin, whose line of symmetry is the x-axis and whose focus is the point $(a, 0)$

For example, the equation $y^2 = 8x$ gives a parabola whose focus is at the point $(2, 0)$.

Example

Find the focus and vertex of the parabola $(y - 1)^2 = 8(x - 2)$ and hence sketch the curve.

Comparing $(y - 1)^2 = 8(x - 2)$

with $\quad Y^2 = 4aX$ whose focus is at $X = a$, $Y = 0$ and vertex is at $X = 0$, $Y = 0$

gives $\quad Y = y - 1$, $X = x - 2$, $a = 2$

$\therefore \qquad$ when $Y = 0$, $y = 1$, when $X = 0$, $x = 2$

Therefore the vertex is at the point $(2, 1)$.

The line of symmetry is $Y = 0$, i.e. $y = 1$,

When $X = 2$, $x = 4$, so the focus is the point $(4, 1)$.

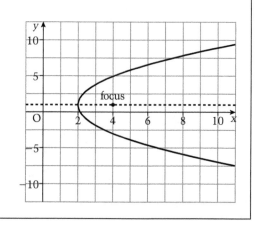

Parametric equations for the standard parabola

The parametric equations of the standard parabola are
$$x = at^2 \text{ and } y = 2at$$

Therefore the coordinates of any point on the parabola can be written as $(at^2, 2at)$.

Using parametric coordinates means that we can find general properties that apply to any point on the parabola.

Example

Find the equation of a chord that passes through the focus of the parabola $x = at^2$ and $y = 2at$ from any point on the parabola.

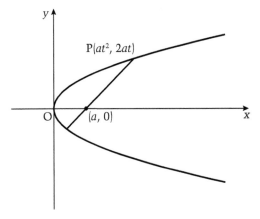

A chord of a curve is a line joining any two points on the curve.

$P(at^2, 2at)$ is any point on the parabola.

The equation of the line through $(a, 0)$ and $(at^2, 2at)$

is $y = \dfrac{2at}{at^2 - a}(x - a)$

$\Rightarrow y(t^2 - 1) = 2t(x - a)$

Example

Find the value of a for which the line $y = 2x - 1$ is a tangent to the parabola $y^2 = 4ax$

Solving the equation of the line and the parabola simultaneously gives
$(2x - 1)^2 = 4ax \Rightarrow 4x^2 - 4x(1 + a) + 1 = 0$

For the line to be a tangent to the curve, this equation must have equal roots, i.e. $16(1 + a)^2 = 16 \Rightarrow 1 + a = \pm 1 \Rightarrow a = -2$

($a = 0$ is not a valid solution to the problem because it gives $y^2 = 0$, which is not a parabola.)

☑ *Exam tip*

This problem highlights the importance of checking that solutions are valid in the context of the problem.

Exercise 2.15

---

1 Find the focus and vertex of the parabola given by:

(a) $(y + 2)^2 = 16x$ (b) $x = 3t^2, y = 6t$

2 Find the coordinates of the points of intersection of the line $y = x - 6$ and the parabola $x = 4t^2, y = 2t$

3 The parametric equations of a curve are $x = 8t^2$, $y = 16t$. Find, in terms of p, the equation of the chord joining the points on the curve where $t = 2$ and $t = p$

2.16 The ellipse

Learning outcomes

- To define the ellipse as a locus
- To find the Cartesian and parametric equations of an ellipse

You need to know

- The meaning of a parameter
- How to find the Cartesian equation of a locus
- How to solve quadratic inequalities

The ellipse

One of the properties of an ellipse discovered by the ancient Greeks is that when a point is constrained so that its distance from a fixed point and a fixed straight line are in a constant ratio which is less than 1, the locus is an ellipse.

The position of the ellipse depends on the position of the fixed point and the fixed line. These are also called the focus and the directrix. The shape of the ellipse depends on the value of the constant ratio; this is called the *eccentricity* of the ellipse and is denoted by e.

(Notice that when $e = 1$, the definition gives a parabola.)

We will find the simplest Cartesian equation for an ellipse. This is when the point $(ae, 0)$ is the focus and the line $x = \dfrac{a}{e}$ is the directrix.

$PN^2 = \left(\dfrac{a}{e} - x\right)^2$ and $PS^2 = (ae - x)^2 + y^2$

Now $ePN = PS$ so $e^2PN^2 = PS^2$

$\therefore \quad e^2 \left(\dfrac{a}{e} - x\right)^2 = (x - ae)^2 - y^2$

$\Rightarrow \quad x^2(1 - e^2) + y^2 = a^2(1 - e^2)$

Replacing $a^2(1 - e^2)$ by b^2 gives $\dfrac{x^2}{a^2} + \dfrac{y^2}{b^2} = 1$

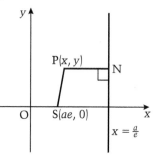

The shape of the curve can be deduced from the equation as follows:

$\dfrac{x^2}{a^2} + \dfrac{y^2}{b^2} = 1 \Rightarrow x^2 = \dfrac{a^2}{b^2}(b^2 - y^2)$ and $x^2 \geqslant 0$

so $b^2 - y^2 \geqslant 0 \Rightarrow (b - y)(b + y) \geqslant 0$

Therefore $-b \leqslant y \leqslant b$, and as

$x = \pm \dfrac{a}{b}\sqrt{b^2 - y^2}$,

the curve is symmetrical about the y-axis.

Solving $\dfrac{x^2}{a^2} + \dfrac{y^2}{b^2} = 1$ for y^2, gives similar

results for x, i.e. $-a \leqslant x \leqslant a$ and the curve is symmetrical about the x-axis.

Also when $x = 0$, $y = \pm b$ and when $y = 0$, $x = \pm a$

The curve is symmetrical about both axes and so we can see that it has two symmetrical foci and directrices.

Did you know?

A property of the ellipse is that the sum of the distances between the foci and between each focus and a point on the ellipse is constant. You can use this to draw an ellipse.

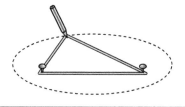

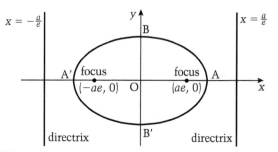

The line AA' is called the major axis and its length is $2a$.

The line through BB' is called the minor axis and its length is $2b$.

$$\frac{x^2}{a^2} + \frac{y^2}{b^2} = 1 \text{ is the Cartesian equation of an ellipse}$$

$$\text{where } b^2 = a^2(1 - e^2),$$

$$a > b \text{ – with foci at } (ae, 0) \text{ and } (-ae, 0),$$

major axis of length $2a$ and minor axis of length $2b$.

Example

Find the eccentricity of the ellipse $\frac{x^2}{9} + \frac{y^2}{4} = 1$

Comparing the given equation with the standard equation gives $a = 3$ and $b = 2$

Using $b^2 = a^2(1 - e^2)$ gives $4 = 9(1 - e^2)$

$\Rightarrow e^2 = \frac{5}{9}$ so $e = \frac{\sqrt{5}}{3}$

Parametric equations of an ellipse

The parametric equations of an ellipse are
$$x = a \cos \theta \text{ and } y = b \sin \theta$$

Therefore the coordinates of any point on the ellipse can be written as $(a \cos \theta, b \sin \theta)$.

Example

Find the length of the chord joining the points A where $\theta = \frac{\pi}{3}$ and B where $\theta = \frac{3\pi}{4}$ on the ellipse $x = a \cos \theta$ and $y = b \sin \theta$

The coordinates of A where $\theta = \frac{\pi}{3}$ are $\left(a \cos \frac{\pi}{3}, b \sin \frac{\pi}{3}\right)$, i.e. $\left(\frac{a}{2}, \frac{b\sqrt{3}}{2}\right)$

The coordinates of B where $\theta = \frac{3\pi}{4}$ are $\left(a \cos \frac{3\pi}{4}, b \sin \frac{3\pi}{4}\right)$, i.e. $\left(-\frac{a\sqrt{2}}{2}, \frac{b\sqrt{2}}{2}\right)$

$$AB^2 = \frac{a^2}{4}(\sqrt{2} + 1)^2 + \frac{b^2}{4}(\sqrt{2} - \sqrt{3})^2$$

$$\Rightarrow AB = \frac{1}{2}\sqrt{a^2(3 + 2\sqrt{2}) + b^2(5 - 2\sqrt{6})}$$

Exercise 2.16

1 Find the lengths of the major and minor axes of the ellipse
   $$\frac{x^2}{25} + \frac{y^2}{9} = 1$$
   Hence sketch the ellipse.

2 (a) Write down the coordinates of F_1 and F_2, the two foci of the ellipse
       $x = 4 \cos \theta, y = 3 \sin \theta$

   (b) Find, in terms of θ, the lengths of F_1P and F_2P where P is the point
       $(4 \cos \theta, 3 \sin \theta)$.

   (c) Hence show that the sum of the lengths F_1P, F_2P and F_1F_2 is constant.

Vectors

A, B and C are three points in space.

AB, BC and AC are displacements. Each has a magnitude (the magnitude of AB is the length of the line segment AB, i.e. 2 m) and a definite direction in space.

The displacement from A to B (written as $\overrightarrow{AB}$) followed by the displacement from B to C is equivalent to the displacement from A to C.

We write this as $\overrightarrow{AB} + \overrightarrow{BC} = \overrightarrow{AC}$

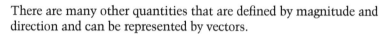

There are many other quantities that are defined by magnitude and direction and can be represented by vectors.

> **A *vector* is a quantity which has both magnitude and a specific direction in space.**

A *scalar* quantity is one that is fully defined by magnitude alone and can be represented by a real number. Length, for example, is a scalar quantity, as the length of a piece of string does not depend on its direction.

A vector can be represented by a straight line segment where the length of the line represents the magnitude and the direction of the line segment represents the direction of the vector.

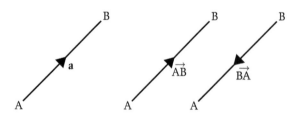

The vector can be denoted by $\overrightarrow{AB}$, where A and B are the end points of the line and the arrow shows the direction, i.e. from A to B. A vector in the opposite direction is denoted by $\overrightarrow{BA}$. The vector can also be denoted by, for example, **a**.

Properties of vectors

The magnitude of a vector **a** is written as $|\mathbf{a}|$ or a, so $|\mathbf{a}|$ is the length of the line representing **a**.

Two vectors are equal if their magnitudes are equal and their directions are equal.

If two vectors **a** and **b** have the same magnitude but opposite directions, then $\mathbf{b} = -\mathbf{a}$

If t is a positive real number, then $t\mathbf{a}$ is in the same direction as **a** but of magnitude $t|\mathbf{a}|$.

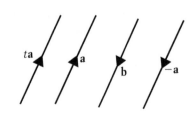

Addition of vectors

If the sides AB and BC represent the vectors **a** and **b**, the third side, AC, represents the vector sum **a** + **b**

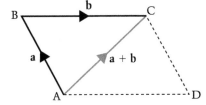
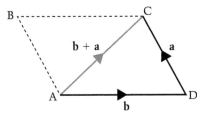

Notice that **a** and **b** follow each other round the triangle (clockwise in the diagram) whereas **a** + **b** is in the opposite sense (anticlockwise in this case).

The order in which **a** and **b** are added does not matter, as the diagrams above show, i.e. **a** + **b** = **b** + **a**

This rule can be extended to cover the addition of as many vectors as we wish to add.

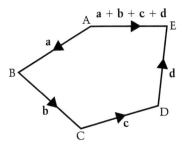

In the diagram the side AE represents the vector sum **a** + **b** + **c** + **d**. Again notice that **a**, **b**, **c** and **d** follow each other round the pentagon in the same sense, but **a** + **b** + **c** + **d** is in the opposite sense.

Note that, although this diagram appears to be two-dimensional, it can equally represent vectors in three dimensions.

Position vectors and displacement vectors

A vector usually has no particular position in space. Such a vector is called a *displacement vector.*

However, some vectors represent the specific position of a point, for example the vector $\overrightarrow{OA}$, where O is a fixed origin, represents the position of the point A relative to O.

$\overrightarrow{OA}$ is called the *position vector* of A. It is unique and cannot be represented by any other line of the same length and direction.

Coordinates in three dimensions

To locate a point in three dimensions we start from a fixed origin, the point O. Any other point can be located by giving its distances from O in each of *three* mutually perpendicular directions. Therefore we need *three* coordinates to locate a point in 3-D.

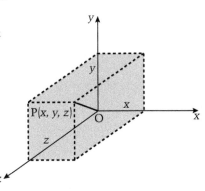

We use the familiar x- and y-axes, together with a third axis Oz. Then any point has coordinates (x, y, z) relative to the origin O.

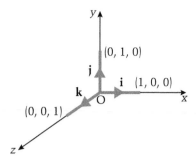

Cartesian unit vectors

A unit vector has a magnitude of one unit.

i is a unit vector in the direction of Ox

j is a unit vector in the direction of Oy

k is a unit vector in the direction of Oz

Therefore the position vector, relative to O, of any point P can be given in terms of **i**, **j** and **k**.

For example, the point P distant

3 units from O in the direction of Ox

4 units from O in the direction of Oy

5 units from O in the direction of Oz

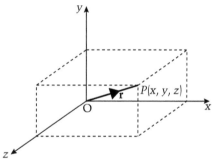

has coordinates $(3, 4, 5)$ and $\overrightarrow{OP} = 3\mathbf{i} + 4\mathbf{j} + 5\mathbf{k}$

This can also be written as $\overrightarrow{OP} = \begin{pmatrix} 3 \\ 4 \\ 5 \end{pmatrix}$

When P is any point with coordinates (x, y, z), then $\mathbf{r} = \overrightarrow{OP}$ is the position vector of P.

Then $\mathbf{r} = x\mathbf{i} + y\mathbf{j} + z\mathbf{k}$

or $\mathbf{r} = \begin{pmatrix} x \\ y \\ z \end{pmatrix}$

Displacement vectors are also given in the same way. For example, the vector $2\mathbf{i} - 3\mathbf{j} + 2\mathbf{k}$ can represent the position vector of the point $P(2, -3, 2)$ but it can equally represent any vector of the same magnitude and direction as OP. Unless we are told that a vector is a position vector, we can assume that it is a displacement vector.

Addition and subtraction of vectors in i, j, k form

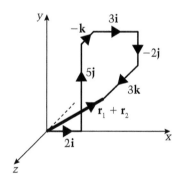

Vectors in **i**, **j**, **k** form can be added and subtracted by adding or subtracting the coefficients of **i**, **j**, and **k** separately.

For example, when $\mathbf{r}_1 = 2\mathbf{i} + 5\mathbf{j} - \mathbf{k}$ and $\mathbf{r}_2 = 3\mathbf{i} - 2\mathbf{j} + 3\mathbf{k}$

then $\mathbf{r}_1 + \mathbf{r}_2 = (2 + 3)\mathbf{i} + (5 - 2)\mathbf{j} + (-1 + 3)\mathbf{k}$

$= 5\mathbf{i} + 3\mathbf{j} + 2\mathbf{k}$

and $\mathbf{r}_1 - \mathbf{r}_2 = (2 - 3)\mathbf{i} + (5 - (-2))\mathbf{j} + (-1 - 3)\mathbf{k}$

$= \mathbf{i} + 7\mathbf{j} - 4\mathbf{k}$

The magnitude of a vector in i, j, k form

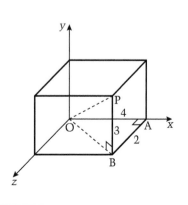

The magnitude of $\mathbf{a} = 4\mathbf{i} + 3\mathbf{j} + 2\mathbf{k}$ is the length of OP where

P is the point $(4, 3, 2)$.

Using Pythagoras' theorem twice gives

$OB^2 = OA^2 + AB^2 = 4^2 + 2^2$

$OP^2 = OB^2 + BP^2 = (4^2 + 2^2) + 3^2 = 4^2 + 3^2 + 2^2$

$\therefore \quad OP = \sqrt{4^2 + 3^2 + 2^2} = \sqrt{29}$

For any vector $\mathbf{r} = x\mathbf{i} + y\mathbf{j} + z\mathbf{k}$, $|\mathbf{r}| = \sqrt{x^2 + y^2 + z^2}$

Parallel vectors

Two vectors if $\mathbf{v}_1$ and $\mathbf{v}_2$ are parallel when $\mathbf{v}_1 = t\mathbf{v}_2$ where $t \in \mathbb{R}$.

For example $3\mathbf{i} + 2\mathbf{j} - \mathbf{k}$ is parallel to $6\mathbf{i} + 4\mathbf{j} - 2\mathbf{k}$ $(t = 2)$
and $3\mathbf{i} + 2\mathbf{j} - \mathbf{k}$ is also parallel to $-3\mathbf{i} - 2\mathbf{j} + \mathbf{k}$ $(t = -1)$

Equal vectors

Vectors $\mathbf{v}_1 = a_1\mathbf{i} + b_1\mathbf{j} + c_1\mathbf{k}$ and $\mathbf{v}_2 = a_2\mathbf{i} + b_2\mathbf{j} + c_2\mathbf{k}$ are equal if and only if $a_1 = a_2$ and $b_1 = b_2$ and $c_1 = c_2$

Example

Determine whether the vector $2\mathbf{i} - 6\mathbf{j} - \mathbf{k}$ is parallel to:

(a) $\mathbf{i} - 2\mathbf{j} - \mathbf{k}$ **(b)** $-\mathbf{i} + 3\mathbf{j} + \frac{1}{2}\mathbf{k}$

(a) $2\mathbf{i} - 6\mathbf{j} - \mathbf{k}$ is not a multiple of $\mathbf{i} - 2\mathbf{j} - \mathbf{k}$, so these vectors are not parallel.

(b) $2\mathbf{i} - 6\mathbf{j} - \mathbf{k} = -2(-\mathbf{i} + 3\mathbf{j} + \frac{1}{2}\mathbf{k})$ so these vectors are parallel.

Example

A is the point $(-1, 3, -2)$ and B is the point $(3, 0, -1)$. Find $|\overrightarrow{AB}|$.

$|\overrightarrow{OA}| = -\mathbf{i} + 3\mathbf{j} - 2\mathbf{k}$ and $|\overrightarrow{OB}| = 3\mathbf{i} - \mathbf{k}$

$\overrightarrow{AB} = \overrightarrow{AO} + \overrightarrow{OB}$

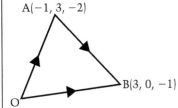

A$(-1, 3, -2)$

B$(3, 0, -1)$

O

When drawing a diagram showing points in three dimensions, do not draw the axes, as they complicate the diagram. However, always include the origin as this gives a reference point.

Remember, the vectors to be added must go round the diagram in the same sense and be in the opposite sense to their sum.

$\overrightarrow{AB} = -(-\mathbf{i} + 3\mathbf{j} - 2\mathbf{k}) + (3\mathbf{i} - \mathbf{k})$

$\qquad = 4\mathbf{i} - 3\mathbf{j} + \mathbf{k}$

$\therefore \quad |\overrightarrow{AB}| = \sqrt{4^2 + 3^2 + 1^2} = \sqrt{26}$

Exercise 2.17

1   P is the point $(1, 4, -2)$.
Give $|\overrightarrow{OP}|$ in $\mathbf{i}$, $\mathbf{j}$, $\mathbf{k}$ form and find the length of OP.

2 $\mathbf{a} = 3\mathbf{i} + 5\mathbf{j} - 2\mathbf{k}$. Write down the vectors:

   (a) $3\mathbf{a}$

   (b) $-\mathbf{a}$

3 $\mathbf{a} = \begin{pmatrix} 1 \\ -2 \\ 0 \end{pmatrix}$ and $\mathbf{b} = \begin{pmatrix} 2 \\ 1 \\ -4 \end{pmatrix}$ are the position vectors of the points

  A and B.

  Show the vectors $\mathbf{a}$ and $\mathbf{b}$ on a diagram, and find the magnitude of $\overrightarrow{BA}$.

Unit vector parallel to a given vector

A unit vector has a magnitude of one unit.

The vector $\mathbf{a} = 2\mathbf{i} - 6\mathbf{j} + 3\mathbf{k}$ is represented by $\overrightarrow{OA}$

$|\mathbf{a}| = \sqrt{6^2 + 2^2 + 3^2} = \sqrt{49} = 7$

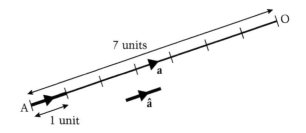

Therefore the unit vector parallel to $\mathbf{a}$ is $\frac{1}{7}$ the magnitude of $\mathbf{a}$,

i.e. the unit vector parallel to $\mathbf{a}$ is $\frac{1}{7}(2\mathbf{i} - 6\mathbf{j} + 3\mathbf{k})$ and is denoted by $\hat{\mathbf{a}}$.

A unit vector in the direction of $\mathbf{v}$ is denoted by $\hat{\mathbf{v}}$ and is given by $\dfrac{\mathbf{v}}{|\mathbf{v}|}$

Example

Find a unit vector in the direction of $\mathbf{v} = \begin{pmatrix} 8 \\ -1 \\ 4 \end{pmatrix}$

$|\mathbf{v}| = \sqrt{64 + 1 + 16} = \sqrt{81} = 9 \quad \therefore \quad \hat{\mathbf{v}} = \frac{1}{9}\begin{pmatrix} 8 \\ -1 \\ 4 \end{pmatrix}$

Exercise 2.18a

1. Find a unit vector in the direction of the vector $\mathbf{i} + 2\mathbf{j} + 2\mathbf{k}$

2. The position vectors of the points A and B are $\begin{pmatrix} 3 \\ 2 \\ 2 \end{pmatrix}$ and $\begin{pmatrix} 1 \\ -4 \\ 0 \end{pmatrix}$ respectively.

   Find a unit vector parallel to $\overrightarrow{AB}$.

Solving problems

To solve a problem in three dimensions, it helps to draw a clear diagram. Mark the origin but do not attempt to draw the axes as these clutter the diagram. Mark all the information on the diagram and draw lines to represent what you need to find. When a diagram is given, copy it so that you can add what you need to find. Remember that any line equal in length and direction to another line can be represented by the same displacement vector.

Example

OABCDEFG is a cube of side 4 units. O is the origin and the unit vectors **i**, **j**, and **k** are parallel to OA, OE and OC respectively. M is the midpoint of the edge DG.

Find, in **i**, **j**, **k** form, the vectors

(a) $\overrightarrow{OA}$ (b) $\overrightarrow{OG}$ (c) $\overrightarrow{OM}$

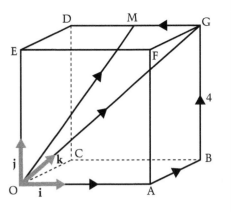

(a) $\overrightarrow{OA} = 4\mathbf{i}$ Each edge of the cube is 4 units long

(b) $\overrightarrow{OG} = \overrightarrow{OA} + \overrightarrow{AB} + \overrightarrow{BG} = 4\mathbf{i} + 4\mathbf{k} + 4\mathbf{j}$

$\qquad = 4\mathbf{i} + 4\mathbf{j} + 4\mathbf{k}$

(c) $\overrightarrow{OM} = \overrightarrow{OG} + \overrightarrow{GM}$

$\quad \overrightarrow{GM} = \frac{1}{2}\overrightarrow{GD}$ and $\overrightarrow{GD} = -\overrightarrow{OA} = -4\mathbf{i}$

$\quad \therefore \quad \overrightarrow{OM} = (4\mathbf{i} + 4\mathbf{j} + 4\mathbf{k}) - \frac{1}{2}(4\mathbf{i})$

$\qquad\qquad = 2\mathbf{i} + 4\mathbf{j} + 4\mathbf{k}$

Example

The position vectors of the points A and B are $2\mathbf{i} + 5\mathbf{j} - 3\mathbf{k}$ and $4\mathbf{i} - 3\mathbf{j} - 2\mathbf{k}$ respectively.

Find the vector of magnitude 5 units in the direction of $\overrightarrow{AB}$.

The vector of magnitude 5 units in the direction of $\overrightarrow{AB}$ is five times the unit vector in the direction of $\overrightarrow{AB}$.

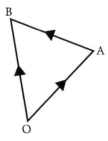

$\overrightarrow{AB} = \overrightarrow{OB} - \overrightarrow{OA}$

$\quad = (4\mathbf{i} - 3\mathbf{j} - 2\mathbf{k}) - (2\mathbf{i} + 5\mathbf{j} - 3\mathbf{k})$

$\quad = 2\mathbf{i} - 8\mathbf{j} + \mathbf{k}$

$|\overrightarrow{AB}| = \sqrt{69}$

$\therefore$ the unit vector in the direction of $\overrightarrow{AB}$ is

$\frac{1}{\sqrt{69}}(2\mathbf{i} - 8\mathbf{j} + \mathbf{k})$ so the required vector is $\frac{5}{\sqrt{69}}(2\mathbf{i} - 8\mathbf{j} + \mathbf{k})$

Exercise 2.18b

1 The position vectors of points A and B are $\overrightarrow{OA} = -\mathbf{i} + 2\mathbf{j} + \mathbf{k}$ and $\overrightarrow{OB} = \mathbf{i} - b\mathbf{j} + \mathbf{k}$ respectively.

(a) Find, in terms of b, the unit vector in the direction of $\overrightarrow{AB}$.

(b) Given that $|\overrightarrow{AB}| = 2$, find the value of b.

2 OABCDE is a right triangular prism with OA = 2 units, OD = 2 units and AB = 4 units. The unit vectors **i**, **j**, and **k** are parallel to OA, OD and OC respectively.

Find the unit vector parallel to $\overrightarrow{DB}$.

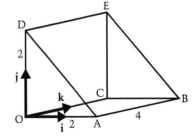

- To define the scalar product of two vectors
- To find the angle between two vectors

- The Cartesian form of a vector
- How to expand the product of two brackets

The angle between two vectors

You can use either the acute angle or the obtuse angle for the angle between two lines, i.e. either α or $\pi - \alpha$

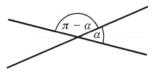

However, the angle between two vectors is defined as the angle between their directions when they both converge or both diverge.

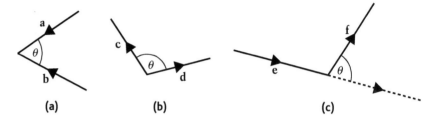

(a)   (b)   (c)

The scalar product

The scalar product of two vectors **a** and **b** is defined as $ab \cos \theta$ where θ is the angle between **a** and **b** and is denoted by **a . b**, i.e.

$$\textbf{a . b} = ab \cos \theta \text{ where } \theta \text{ is the angle between a and b}$$

Parallel vectors

When **a** and **b** are parallel, then

either $\textbf{a . b} = ab \cos 0$ or $\textbf{a . b} = ab \cos \pi$

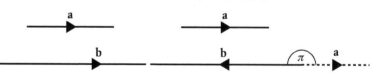

Now $\cos 0 = 1$ and $\cos \pi = -1$, therefore

for   parallel vectors in the same direction a . b = ab
and for   parallel vectors in opposite directions a . b = $-ab$

In the special case when $\textbf{a} = \textbf{b}$, $\textbf{a . b} = \textbf{a . a} = a^2$

For the unit vectors, **i**, **j** and **k**,

$$\textbf{i . i} = \textbf{j . j} = \textbf{k . k} = 1$$

Perpendicular vectors

When **a** and **b** are perpendicular, then $\textbf{\textit{a . b}} = ab \cos \dfrac{\pi}{2}$
but $\cos \dfrac{\pi}{2} = 0$, therefore

for perpendicular vectors a and b, a . b = 0

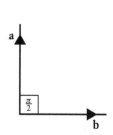

In particular, for the unit vectors **i**, **j** and **k**,

$$\textbf{i . j} = \textbf{i . k} = \textbf{j . k} = 0$$

The scalar product of vectors in Cartesian form

When $\mathbf{a} = x_1\mathbf{i} + y_1\mathbf{j} + z_1\mathbf{k}$ and $\mathbf{b} = x_2\mathbf{i} + y_2\mathbf{j} + z_2\mathbf{k}$,
$\mathbf{a} \cdot \mathbf{b} = (x_1\mathbf{i} + y_1\mathbf{j} + z_1\mathbf{k}) \cdot (x_2\mathbf{i} + y_2\mathbf{j} + z_2\mathbf{k})$

Expanding these brackets gives terms involving $\mathbf{i} \cdot \mathbf{i}$, $\mathbf{j} \cdot \mathbf{j}$ and $\mathbf{k} \cdot \mathbf{k}$, together with terms involving $\mathbf{i} \cdot \mathbf{j}$, $\mathbf{i} \cdot \mathbf{k}$ and $\mathbf{j} \cdot \mathbf{k}$ that are all zero as they involve the scalar product of perpendicular vectors.

Now $\mathbf{i} \cdot \mathbf{i} = \mathbf{j} \cdot \mathbf{j} = \mathbf{k} \cdot \mathbf{k} = 1$,
therefore $\mathbf{a} \cdot \mathbf{b} = x_1x_2 + y_1y_2 + z_1z_2$

i.e. $\qquad (x_1\mathbf{i} + y_1\mathbf{j} + z_1\mathbf{k}) \cdot (x_2\mathbf{i} + y_2\mathbf{j} + z_2\mathbf{k}) = x_1x_2 + y_1y_2 + z_1z_2$

For example $(3\mathbf{i} - 2\mathbf{j} + \mathbf{k}) \cdot (2\mathbf{i} + 5\mathbf{j} - 2\mathbf{k}) = (3)(2) + (-2)(5) + (1)(-2) = -6$

Example

Find the value of a for which the vectors $\begin{pmatrix} 2 \\ -1 \\ 1 \end{pmatrix}$ and $\begin{pmatrix} 2 \\ a \\ -2 \end{pmatrix}$ are perpendicular.

$\begin{pmatrix} 2 \\ -1 \\ 1 \end{pmatrix} \cdot \begin{pmatrix} 2 \\ a \\ -2 \end{pmatrix} = 4 - a - 2$

The vectors are perpendicular when $4 - a - 2 = 0$
$\therefore \quad a = 2$

Example

The position vectors of points A and B are $\overrightarrow{OA} = \mathbf{i} + 2\mathbf{j} + 3\mathbf{k}$ and $\overrightarrow{OB} = \mathbf{i} - 3\mathbf{j} + 2\mathbf{k}$, respectively.
Find the angle between $\overrightarrow{OA}$ and $\overrightarrow{OB}$.

$|\overrightarrow{OA}| = \sqrt{14}$ and $|\overrightarrow{OB}| = \sqrt{14}$,
$\overrightarrow{OA} \cdot \overrightarrow{OB} = 1 - 6 + 6 = 1$
$\therefore \quad |\overrightarrow{OA}| \times |\overrightarrow{OB}| \times \cos \angle AOB = 1$
so $\cos \angle AOB = \dfrac{1}{|\overrightarrow{OA}||\overrightarrow{OB}|} = \dfrac{1}{14}$
$\Rightarrow \angle AOB = 1.50\,\text{rad}$

Exercise 2.19

1 $\mathbf{a} = 4\mathbf{i} - 3\mathbf{j} + 5\mathbf{k}$ and $\mathbf{b} = 2\mathbf{i} - 2\mathbf{j} - 4\mathbf{k}$

Find $\mathbf{a} \cdot \mathbf{b}$ and the angle between $\mathbf{a}$ and $\mathbf{b}$.

2 Show that $\mathbf{i} + 7\mathbf{j} + 3\mathbf{k}$ is perpendicular to both $\mathbf{i} - \mathbf{j} + 2\mathbf{k}$ and $2\mathbf{i} + \mathbf{j} - 3\mathbf{k}$

Learning outcomes

- To define the vector, Cartesian and parametric equations of a line

You need to know

- The difference between a position vector and a displacement vector
- How to add and subtract vectors
- The Cartesian form of a vector in three dimensions
- The scalar product of two vectors

Straight lines in three dimensions

A straight line is uniquely located in space if:

- it is parallel to a given vector, i.e. it has a known direction, and passes through a fixed point, or

- it passes through two fixed points.

The vector equation of a line

The line L passes through the point A whose position vector is **a** and is parallel to the vector **b**.

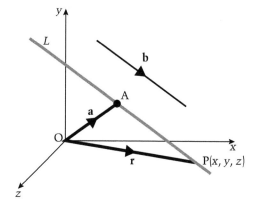

$P(x, y, z)$ is any point on the line.

If **r** is the position vector of P, i.e. $\mathbf{r} = \overrightarrow{OP}$, then $\overrightarrow{AP} = t\mathbf{b}$, where t can take any real value.

Now $\overrightarrow{OP} = \overrightarrow{OA} + \overrightarrow{AP}$,

i.e. $\mathbf{r} = \mathbf{a} + \lambda\mathbf{b}$

and for any value of λ, this equation gives a point on the line.

> $\mathbf{r} = \mathbf{a} + \lambda\mathbf{b}$ is called the vector equation of the line
> where **a** is the position vector of a point on the line
> and **b** is a vector parallel to the line.

Now **a** is the position vector of *any* point on the line so it can have many different values. This means that **the vector equation of a line is not unique** although the line is unique.

For example, the line whose vector equation is $\mathbf{r} = \mathbf{i} - 2\mathbf{j} + \mathbf{k} + \lambda(3\mathbf{i} - \mathbf{k})$ is parallel to $3\mathbf{i} - \mathbf{k}$ and $\mathbf{i} - 2\mathbf{j} + \mathbf{k}$ is the position vector of a point on the line.

Example

(a) Write down the position vector of two points on the line whose vector equation is

$$\mathbf{r} = 5\mathbf{i} - \mathbf{j} + 2\mathbf{k} + \lambda(3\mathbf{i} + 4\mathbf{j} - 6\mathbf{k})$$

(b) Determine whether the vector $-6\mathbf{i} - 8\mathbf{j} + 12\mathbf{k}$ is parallel to the line.

(c) Show that the vector $2\mathbf{i} + 3\mathbf{j} + 3\mathbf{k}$ is perpendicular to the line.

(a) Comparing $\mathbf{r} = 5\mathbf{i} - \mathbf{j} + 2\mathbf{k} + \lambda(3\mathbf{i} + 4\mathbf{j} - 6\mathbf{k})$ with $\mathbf{r} = \mathbf{a} + \lambda\mathbf{b}$ gives $\mathbf{a} = 5\mathbf{i} - \mathbf{j} + 2\mathbf{k}$ so this is one point on the line.

Giving λ any value gives another point on the line, so taking $\lambda = 1$ gives $\mathbf{r} = 8\mathbf{i} + 3\mathbf{j} - 4\mathbf{k}$

Therefore $5\mathbf{i} - \mathbf{j} + 2\mathbf{k}$ and $8\mathbf{i} + 3\mathbf{j} - 4\mathbf{k}$ are the position vectors of two points on the line.

(b) Comparing $\mathbf{r} = 5\mathbf{i} - \mathbf{j} + 2\mathbf{k} + \lambda(3\mathbf{i} + 4\mathbf{j} - 6\mathbf{k})$ with $\mathbf{r} = \mathbf{a} + \lambda\mathbf{b}$ shows that $3\mathbf{i} + 4\mathbf{j} - 6\mathbf{k}$ is parallel to the line.

$-6\mathbf{i} - 8\mathbf{j} + 12\mathbf{k} = -2(3\mathbf{i} + 4\mathbf{j} - 6\mathbf{k})$, therefore $-6\mathbf{i} - 8\mathbf{j} + 12\mathbf{k}$ is also parallel to the line.

(c) The vector $3\mathbf{i} + 4\mathbf{j} - 6\mathbf{k}$ is parallel to the line, so if the scalar product of $3\mathbf{i} + 4\mathbf{j} - 6\mathbf{k}$ and $2\mathbf{i} + 3\mathbf{j} + 3\mathbf{k}$ is zero, then the vectors are perpendicular.

$(3\mathbf{i} + 4\mathbf{j} - 6\mathbf{k}) \cdot (2\mathbf{i} + 3\mathbf{j} + 3\mathbf{k}) = 6 + 12 - 18 = 0$

Therefore $2\mathbf{i} + 3\mathbf{j} + 3\mathbf{k}$ is perpendicular to the line.

Parametric equations of a line

If $P(x, y, z)$ is any point on the line $\mathbf{r} = 3\mathbf{i} + 2\mathbf{j} + 5\mathbf{k} + \lambda(8\mathbf{i} + 4\mathbf{j} - 6\mathbf{k})$

then $\qquad x\mathbf{i} + y\mathbf{j} + z\mathbf{k} = 3\mathbf{i} + 2\mathbf{j} + 5\mathbf{k} + \lambda(8\mathbf{i} + 4\mathbf{j} - 6\mathbf{k})$

Equating the coefficients of $\mathbf{i}$, $\mathbf{j}$ and $\mathbf{k}$ gives

$$x = 3 + 8\lambda, \, y = 2 + 4\lambda, \, z = 5 - 6\lambda$$

These equations give the coordinates of any point on the line in terms of the parameter λ and are called the parametric equations of the line.

Now $3\mathbf{i} + 2\mathbf{j} + 5\mathbf{k}$ is the position vector of *any* point on the line so it can have many different values. This means that **the parametric equations of a line (like the vector equation) are not unique** although the line is unique.

The parametric equations of any line can be found in the same way, i.e. if $P(x, y, z)$ is any point on the line $\mathbf{r} = x_1\mathbf{i} + y_1\mathbf{j} + z_1\mathbf{k} + \lambda(a\mathbf{i} + b\mathbf{j} + c\mathbf{k})$

then $\qquad x\mathbf{i} + y\mathbf{j} + z\mathbf{k} = x_1\mathbf{i} + y_1\mathbf{j} + z_1\mathbf{k} + \lambda(a\mathbf{i} + b\mathbf{j} + c\mathbf{k})$

Therefore $x = x_1 + \lambda a$, $y = y_1 + \lambda b$, $z = z_1 + \lambda c$ are the parametric equations of a line where (x_1, y_1, x_1) is a point on the line and $a\mathbf{i} + b\mathbf{j} + c\mathbf{k}$ is a vector parallel to the line.

For example, the equations

$$x = -1 + 2\lambda, \, y = 3 - 7\lambda, \, z = 1 + 4\lambda$$

are the equations of a line where $(-1, 3, 1)$ is a point on the line and $2\mathbf{i} - 7\mathbf{j} + 4\mathbf{k}$ is parallel to the line.

Example

Find the parametric equations of the line through the points A(1, 1, 4) and B(0, −1, 2).

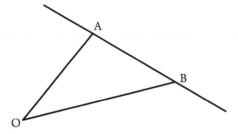

$\overrightarrow{AB}$ is parallel to the line and $\overrightarrow{AB} = \overrightarrow{OB} - \overrightarrow{OA}$

Therefore $\overrightarrow{AB} = -\mathbf{j} + 2\mathbf{k} - (\mathbf{i} + \mathbf{j} + 4\mathbf{k}) = -\mathbf{i} - 2\mathbf{j} - 2\mathbf{k}$

Using A(1, 1, 4) as a point on the line and comparing with
$x = x_1 + \lambda a,\ y = y_1 + \lambda b,\ z = z_1 + \lambda c$ gives $x_1 = 1, y_1 = 1, z_1 = 4$
and $a = -1, b = -2, c = -2$

Therefore the parametric equations of the line are

$x = 1 - \lambda,\ y = 1 - 2\lambda,\ z = 4 - 2\lambda$

Cartesian equations of a line

Starting with the parametric equations of a line, $x = x_1 + \lambda a$, $y = y_1 + \lambda b$, $z = z_1 + \lambda c$ and solving each for λ gives the Cartesian equations of a line, i.e.

$$\frac{x - x_1}{a} = \frac{y - y_1}{b} = \frac{z - z_1}{c}\ (= \lambda)$$

where (x_1, y_1, x_1) is a point on the line
and $a\mathbf{i} + b\mathbf{j} + c\mathbf{k}$ is a vector parallel to the line.

Using the equations of a line

Any of the three forms described above can be used to describe the equation of a line and in each of them, you can 'read' the coordinates of a point on the line and a vector that is parallel to the line.

Example

State whether the lines with equations
$\mathbf{r} = 2\mathbf{i} - 3\mathbf{j} + 2\mathbf{k} + \lambda(\mathbf{i} - \mathbf{j} + 4\mathbf{k})$ and
$\mathbf{r} = (3 - \mu)\mathbf{i} - (3 - \mu)\mathbf{j} + (2 - 4\mu)\mathbf{k}$ are parallel.

To determine whether the lines are parallel we need to find a vector in the direction of each line. We can 'read' this from the equation of the first line but the equation of the second line needs rearranging first.

$\mathbf{r} = 2\mathbf{i} - 3\mathbf{j} + 2\mathbf{k} + \lambda(\mathbf{i} - \mathbf{j} + 4\mathbf{k})$ is parallel to the vector $(\mathbf{i} - \mathbf{j} + 4\mathbf{k})$

$\mathbf{r} = (3 - \mu)\mathbf{i} - (3 - \mu)\mathbf{j} + (2 - 4\mu)\mathbf{k} \Rightarrow \mathbf{r} = 3\mathbf{i} - 3\mathbf{j} + 2\mathbf{k} + \mu(-\mathbf{i} + \mathbf{j} - 4\mathbf{k})$,

so this line is parallel to the vector $(-\mathbf{i} + \mathbf{j} - 4\mathbf{k})$

$\mathbf{i} - \mathbf{j} + 4\mathbf{k} = -(-\mathbf{i} + \mathbf{j} - 4\mathbf{k})$ therefore the lines are parallel.

Example

The point C is the midpoint of the line segment joining A(3, 2, −2) and B(2, 1, 5). D is the point (1, −4, 1).

Find the Cartesian equations of the line through C and D.

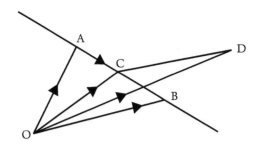

To find the Cartesian equations of the line through C and D, we need a vector parallel to CD.

C is the point with position vector $\overrightarrow{OA} + \overrightarrow{AC}$

and $\overrightarrow{AC} = \frac{1}{2}\overrightarrow{AB} = \frac{1}{2}(\overrightarrow{OB} - \overrightarrow{OA})$

$$= \frac{1}{2}\{(2\mathbf{i} + \mathbf{j} + 5\mathbf{k}) - (3\mathbf{i} + 2\mathbf{j} - 2\mathbf{k})\}$$

$$= \frac{1}{2}(-\mathbf{i} - \mathbf{j} + 7\mathbf{k})$$

$\therefore \overrightarrow{OC} = (3\mathbf{i} + 2\mathbf{j} - 2\mathbf{k}) + \frac{1}{2}(-\mathbf{i} - \mathbf{j} + 7\mathbf{k})$

$$= \frac{1}{2}(5\mathbf{i} + 3\mathbf{j} + 3\mathbf{k})$$

$\overrightarrow{CD} = \overrightarrow{OD} - \overrightarrow{OC}$

$$= (\mathbf{i} - 4\mathbf{j} + \mathbf{k}) - \frac{1}{2}(5\mathbf{i} + 3\mathbf{j} + 3\mathbf{k}) = \frac{1}{2}(-3\mathbf{i} - 11\mathbf{j} - \mathbf{k})$$

Using D(1, −4, 1) as a point on the line and $3\mathbf{i} + 11\mathbf{j} + \mathbf{k}$ as a vector parallel to CD gives the Cartesian equations as

$$\frac{x - 1}{3} = \frac{y + 4}{11} = \frac{z - 1}{1}$$

Example

The line L_1 has equations $x = 2 - 3\lambda$, $y = 1 + \lambda$, $z = 5 - 2\lambda$ and the line L_2 has equations $x = 1 + 4\mu$, $y = 4 - 3\mu$, $z = -1 + \mu$.
Find the angle between L_1 and L_2.

L_1 is parallel to the vector $-3\mathbf{i} + \mathbf{j} - 2\mathbf{k}$ and L_2 is parallel to the vector $4\mathbf{i} - 3\mathbf{j} + \mathbf{k}$.

If θ is the angle between L_1 and L_2, using the scalar product,

$(-3\mathbf{i} + \mathbf{j} - 2\mathbf{k}) \cdot (4\mathbf{i} - 3\mathbf{j} + \mathbf{k}) = |-3\mathbf{i} + \mathbf{j} - 2\mathbf{k}|\ |4\mathbf{i} - 3\mathbf{j} + \mathbf{k}| \cos \theta$

$\Rightarrow -12 - 3 - 2 = (\sqrt{9 + 1 + 4})(\sqrt{16 + 9 + 1}) \cos \theta$

$\Rightarrow \qquad \cos \theta = -\dfrac{17}{\sqrt{14}\sqrt{26}}$

$\therefore \quad \theta = 2.67$ rad   This is the obtuse angle between the lines; the acute angle is 0.471 rad

Exercise 2.20

1 A is the point (1, 2, 2) and B is the point (2, 0, 5). Find vector and parametric equations for the line through A and B.

2 The line l_1 has equation $\mathbf{r} = \begin{pmatrix} 2 \\ -1 \\ 5 \end{pmatrix} + \lambda \begin{pmatrix} 4 \\ 2 \\ -1 \end{pmatrix}$ and the line l_2 has

equation $\mathbf{r} = \begin{pmatrix} 5 \\ 1 \\ 1 \end{pmatrix} + \mu \begin{pmatrix} 3 \\ a \\ 8 \end{pmatrix}$

Given that l_1 and l_2 are perpendicular, find the value of a.

2.21 Pairs of lines

Learning outcomes

- To determine whether two lines in three dimensions are parallel, intersecting or skew

You need to know

- The different forms for the equations of a line in three dimensions
- The condition for vectors to be parallel
- How to solve a pair of simultaneous equations

Pairs of lines in space

Two lines in space may be parallel or not parallel, in which case they may intersect or they may not.

A pair of non-parallel lines that do not intersect are called **skew**.

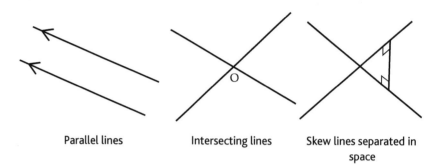

Parallel lines   Intersecting lines   Skew lines separated in space

Parallel lines

It is easy to tell whether two lines are parallel because you can 'read' the vectors that are parallel to each line from their equations.

For example, the lines
$\mathbf{r} = 2\mathbf{i} - \mathbf{k} + \lambda(2\mathbf{i} - \mathbf{j} + \mathbf{k})$ and $\mathbf{r} = \mathbf{j} + 2\mathbf{k} + \mu(4\mathbf{i} - 2\mathbf{j} + 2\mathbf{k})$
are parallel because the vectors parallel to the lines,
i.e. $2\mathbf{i} - \mathbf{j} + \mathbf{k}$ and $4\mathbf{i} - 2\mathbf{j} + 2\mathbf{k}$ $(= 2(2\mathbf{i} - \mathbf{j} + \mathbf{k}))$, are parallel.

Non-parallel lines

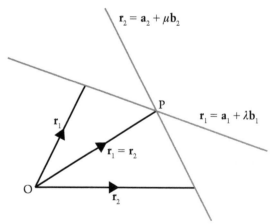

Two lines whose vector equations are $\mathbf{r}_1 = \mathbf{a}_1 + \lambda\mathbf{b}_1$ and $\mathbf{r}_2 = \mathbf{a}_2 + \mu\mathbf{b}_2$ intersect if values of λ and μ can be found for which $\mathbf{r}_1 = \mathbf{r}_2$

If no such values can be found then the lines are skew.

The parametric equations of lines are easiest to work with when determining whether two lines intersect or are skew.

Example

Show that the lines $x = 1 + \lambda$, $y = -1 - \lambda$, $z = 3 + \lambda$ and
$x = 2 + 2\mu$, $y = 4 + \mu$, $z = 6 + 3\mu$ intersect and find the coordinates
of their point of intersection.

If the lines intersect then equating the values of x gives
$$1 + \lambda = 2 + 2\mu \quad [1]$$
and equating the values of y gives $-1 - \lambda = 4 + \mu \quad [2]$

Solving these equations gives $\lambda = -3$ and $\mu = -2$

When $\lambda = -3$, $x = -2$, $y = 2$ and $z = 0$ (first line)

When $\mu = -2$, $x = -2$, $y = 2$ and $z = 0$ (second line)

Therefore these values of λ and μ give the same point on each line, so
the lines intersect at the point $(-2, 2, 0)$.

Example

Determine whether the lines $\mathbf{r}_1 = \mathbf{i} + \mathbf{k} + \lambda(\mathbf{i} + 3\mathbf{j} - \mathbf{k})$ and
$\mathbf{r}_2 = 2\mathbf{i} + 3\mathbf{j} + \mathbf{k} + \mu(4\mathbf{i} - \mathbf{j} - 5\mathbf{k})$ are parallel, intersect or are skew.

The lines are not parallel because $\mathbf{i} + 3\mathbf{j} - \mathbf{k}$ and $4\mathbf{i} - \mathbf{j} - 4\mathbf{k}$
are not parallel.

Writing the equations in parametric form:
$$x_1 = 1 + \lambda, \qquad y_1 = 3\lambda, \qquad z_1 = 1 - \lambda$$
$$x_2 = 2 + 4\mu, \qquad y_2 = 3 - \mu, \qquad z_2 = 1 - 5\mu$$

Equating x_1 and x_2 gives $1 + \lambda = 2 + 4\mu \quad [1]$

Equating y_1 and y_2 gives $3\lambda = 3 - \mu \quad [2]$

Solving [1] and [2] gives $\lambda = 1$ and $\mu = 0$

With these values $z_1 = 0$ and $z_2 = 1$

These values are not equal, therefore the lines do not intersect and are
skew.

Exercise 2.21

1 Show that the lines $\mathbf{r} = \begin{pmatrix} 2 \\ 1 \\ 2 \end{pmatrix} + \lambda \begin{pmatrix} -2 \\ 3 \\ 2 \end{pmatrix}$ and $\mathbf{r} = \begin{pmatrix} 5 \\ 1 \\ 0 \end{pmatrix} + \mu \begin{pmatrix} -1 \\ 0 \\ 2 \end{pmatrix}$ are skew.

2 Two lines which intersect have equations
$$\mathbf{r}_1 = 2\mathbf{i} + 9\mathbf{j} + 13\mathbf{k} + \lambda(\mathbf{i} + 2\mathbf{j} + 3\mathbf{k}) \text{ and}$$
$$\mathbf{r}_2 = a\mathbf{i} + 7\mathbf{j} - 2\mathbf{k} + \mu(-\mathbf{i} + 2\mathbf{j} - 3\mathbf{k})$$
Find: **(a)** the value of a

(b) the position vector of the point of intersection

(c) the angle between the lines.

Learning outcomes

- To determine the vector and Cartesian equation of a plane

You need to know

- How to add and subtract vectors in three dimensions
- The scalar product of two vectors
- The equations of a line in three dimensions
- The meaning of a unit vector

Defining a plane

There are several ways to define a unique plane, for example:

(a) one and only one plane can be drawn through three non-collinear points, therefore three given points specify a unique plane

(b) one and only one plane can be drawn to contain two intersecting lines, therefore two given intersecting lines specify a unique plane

(c) one and only one plane can be drawn perpendicular to a given direction at a given distance from the origin, therefore the normal to a plane and the distance of the plane from the origin specify a unique plane

(d) one and only one plane can be drawn through a given point and perpendicular to a given direction, therefore a point on the plane and a normal to the plane specify a unique plane.

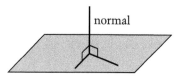

(A **normal** to a plane is any line perpendicular to the plane. A normal is therefore perpendicular to any line in the plane.)

The vector equation of a plane

We use the definition of a plane given in **(d)** to derive the vector equation of a plane.

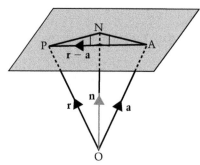

A is a point on the plane and $\overrightarrow{OA} = \mathbf{a}$

The vector $\mathbf{n}$ is perpendicular to the plane.

P is any point on the plane and $\overrightarrow{OP} = \mathbf{r}$

AP is a line in the plane, and is therefore perpendicular to $\mathbf{n}$.

$\overrightarrow{AP} = \mathbf{r} - \mathbf{a}$, therefore the scalar product of $\mathbf{r} - \mathbf{a}$ and $\mathbf{n}$ is zero,

$$\text{i.e. } (\mathbf{r} - \mathbf{a}) \cdot \mathbf{n} = 0$$

This is called the vector equation of a plane.

The vector equation of a plane can be written in another form.

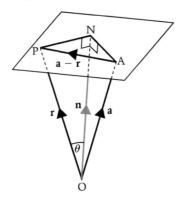

In the diagram, ON is the distance of the plane from the origin.

If ON $= d$, then in triangle OPN, $d = $ OP $\cos\theta$

But $\mathbf{r} . \hat{\mathbf{n}} = $ OP $\cos\theta = d$

> **Therefore $\mathbf{r} . \hat{\mathbf{n}} = d$ is the vector equation of a plane that is perpendicular to the unit vector $\hat{\mathbf{n}}$ and distant d from the origin.**
>
> **This equation can be multiplied by any scalar.**
> **Therefore an equation of the form $\mathbf{r} . \mathbf{n} = D$ represents a plane**
> **perpendicular to $\mathbf{n}$ and distant $\dfrac{D}{\hat{\mathbf{n}}}$ from the origin.**

Example

The vector equation of a plane is $\mathbf{r} . (2\mathbf{i} + \mathbf{j} + 2\mathbf{k}) = 12$.
Find the distance of the plane from the origin.

$2\mathbf{i} + \mathbf{j} + 2\mathbf{k}$ is a vector perpendicular to the plane.

Dividing both sides by $|2\mathbf{i} + \mathbf{j} + 2\mathbf{k}|$ converts the equation to the form $\mathbf{r} . \hat{\mathbf{n}} = d$

$|2\mathbf{i} + \mathbf{j} + 2\mathbf{k}| = 3$, so the equation becomes $\mathbf{r} . \frac{1}{3}(2\mathbf{i} + \mathbf{j} + 2\mathbf{k}) = 4$

Therefore the plane is 4 units from the origin.

Example

(a) Find the equation of the plane that is perpendicular to the
  vector $\begin{pmatrix} 4 \\ 4 \\ -1 \end{pmatrix}$ and contains the point $\begin{pmatrix} 2 \\ 5 \\ 3 \end{pmatrix}$.

(b) Find the distance of the plane from the origin.

(a) Using the form $(\mathbf{r} - \mathbf{a}) . \mathbf{n} = 0 \Rightarrow \mathbf{r} . \mathbf{n} - \mathbf{a} . \mathbf{n} = 0$ gives

$$\mathbf{r} . \begin{pmatrix} 4 \\ 4 \\ -1 \end{pmatrix} - \begin{pmatrix} 2 \\ 5 \\ 3 \end{pmatrix} . \begin{pmatrix} 4 \\ 4 \\ -1 \end{pmatrix} = 0$$

$$\mathbf{r} . \begin{pmatrix} 4 \\ 4 \\ -1 \end{pmatrix} - 25 = 0, \quad \text{i.e.} \quad \mathbf{r} . \begin{pmatrix} 4 \\ 4 \\ -1 \end{pmatrix} = 25$$

(b) The distance of the plane from the origin $= \dfrac{25}{\sqrt{4^2 + 4^2 + (-1)^2}} = \dfrac{25}{\sqrt{33}}$

Example

Show that the line whose equation is $\mathbf{r} = \begin{pmatrix} 3 \\ -1 \\ 2 \end{pmatrix} + \lambda\begin{pmatrix} 2 \\ 2 \\ 5 \end{pmatrix}$ is parallel to

the plane $\mathbf{r} \cdot \begin{pmatrix} 8 \\ 2 \\ -4 \end{pmatrix} = 5$

If the line is parallel to the plane, then it is perpendicular to the normal to the plane.

The line is parallel to $\begin{pmatrix} 2 \\ 2 \\ 5 \end{pmatrix}$ and the plane is perpendicular to $\begin{pmatrix} 8 \\ 2 \\ -4 \end{pmatrix}$.

Now $\begin{pmatrix} 2 \\ 2 \\ 5 \end{pmatrix} \cdot \begin{pmatrix} 8 \\ 2 \\ -4 \end{pmatrix} = 0$, therefore the line is parallel to the plane.

The Cartesian equation of a plane

The point P(x, y, z) is any point on the plane that is perpendicular to the unit vector where

$$\hat{\mathbf{n}} = l\mathbf{i} + m\mathbf{j} + n\mathbf{k}$$

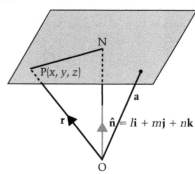

Using $\mathbf{r} \cdot \hat{\mathbf{n}} = d$ gives

$$(x\mathbf{i} + y\mathbf{j} + z\mathbf{k}) \cdot (l\mathbf{i} + m\mathbf{j} + n\mathbf{k}) = d$$
$$\Rightarrow \quad lx + my + nz = d$$

Therefore

$lx + my + nz = d$ **is the Cartesian equation of a plane**
where d is the distance of the plane from the origin
and $l\mathbf{i} + m\mathbf{j} + n\mathbf{k}$ is a unit vector perpendicular to the plane.

Multiplying this equation by a constant gives the more general form of the Cartesian equation,

i.e. $\qquad\qquad ax + by + cz = D$

Then $a\mathbf{i} + b\mathbf{j} + c\mathbf{k}$ is a vector perpendicular to the plane and the distance of the plane from the origin is given by

$$\frac{D}{\sqrt{a^2 + b^2 + c^2}}.$$

For example, the equation $3x - 2y + 6z = 21$ represents a plane where $3\mathbf{i} - 2\mathbf{j} + 6\mathbf{k}$ is perpendicular to the plane and the distance of the plane from the origin is given by

$$\frac{21}{\sqrt{3^2 + 2^2 + 6^2}} = \frac{21}{7} = 3$$

Example

(a) Find the Cartesian equation of the plane that contains the points A(1, 5, −2), B(1, 1, 1) and C(5, −1, 4).

(b) Hence write down a vector that is perpendicular to the plane.

(a) The Cartesian equation of a plane is $ax + by + cz = D$

(1, 5, −2) satisfies the equation of the plane,

therefore $\quad\quad\quad a + 5b − 2c = D \quad\quad$ [1]

(1, 1, 1) also satisfies the equation of the plane,

therefore $\quad\quad\quad a + b + c = D \quad\quad$ [2]

similarly, using C $\quad 5a − b + 4c = D \quad\quad$ [3]

These three equations are enough to find a, b and c in terms of D. This is all we need because any multiple of $ax + by + cz = D$ is also the equation of the plane.

[1] − [2] gives $\quad\quad 4b − 3c = 0 \quad\quad$ [4]

5[2] − [3] gives $\quad\quad 6b + c = 4D \quad\quad$ [5]

3[5] + [4] gives $\quad\quad b = \dfrac{6D}{11}$

Then [4] gives $\quad\quad c = \dfrac{8D}{11} \quad$ and [2] gives $\quad a = −\dfrac{3D}{11}$

Therefore the equation of the plane is

$$\dfrac{-3D}{11}x + \dfrac{6D}{11}y + \dfrac{8D}{11}z = D$$

$\Rightarrow \quad −3x + 6y + 8z = 11$

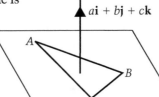

(b) $a\mathbf{i} + b\mathbf{j} + c\mathbf{k}$ is perpendicular to the plane $ax + by + cz = D$

Therefore $−3\mathbf{i} + 6\mathbf{j} + 8\mathbf{k}$ is perpendicular to the plane.

Exercise 2.22

1 Find, in the form $\mathbf{r} \cdot \mathbf{n} = D$, the equation of the plane that contains the point (2, 5, −1) and is perpendicular to the vector $3\mathbf{i} − \mathbf{j} + \mathbf{k}$. Hence write down the distance of the plane from the origin.

2 Find the Cartesian equation of the plane that contains the point (1, 0 −1) and is perpendicular to the vector $\mathbf{i} − 2\mathbf{j} − \mathbf{k}$.

3 (a) Show that the line L with equations $x = 2 − \lambda$, $y = 1 + \lambda$, $z = 3 + 2\lambda$ is parallel to the plane P whose vector equation is $\mathbf{r} \cdot (−3\mathbf{i} + \mathbf{j} − 2\mathbf{k}) = 6$

(b) Show that the point (0, 8, 1) is contained in the plane. Hence find the equation of the line parallel to L that is contained in the plane P.

Section 2   Practice questions

1 Give exact values for:

   (a) $\cos\left(\dfrac{4\pi}{3}\right)$ (c) $\sin\left(\dfrac{7\pi}{4}\right)$

   (b) $\tan\left(\dfrac{7\pi}{4}\right)$ (d) $\operatorname{cosec}\left(\dfrac{5\pi}{3}\right)$

2 Give the minimum value of

$$\dfrac{3}{2\sin\left(\theta-\frac{\pi}{3}\right)}$$

3 Given $\sin A = \frac{5}{13}$ and that A is acute, find the value of:

   (a) $\sin 2A$ (b) $\cot A$

4 (a) Prove the identity

$$\tan^2\theta - \cot^2\theta \equiv -\dfrac{4\cot 2\theta}{\sin 2\theta}$$

   (b) Hence find the solution of the equation

$$\tan^2\theta - \cot^2\theta = 4$$

     for values of θ between $-\pi$ and π.

5 Prove that

$$\cos A\cos 5A - \sin A\sin 5A \equiv \cos 6A$$

6 Express $\cos x - \sqrt{3}\sin x$ in the form $r\cos(x + \alpha)$.

Hence find the maximum and minimum values of $\cos x - \sqrt{3}\sin x$ and the values of x at which they occur in the range $0 \leqslant x \leqslant 2\pi$

7 Show that

$$(2\sin\theta + 5\cos\theta)^2 \leqslant 29$$

for all values of θ.

8 Prove the identity

$$\sin^2 2x\cot^2 x - \sin^2 2x\tan^2 x \equiv 4\cos 2x$$

9 Solve the equation

$$\sin 3\theta - \sin 2\theta + \sin\theta = 0$$

for values of θ between 0 and 2π.

10 Find the general solution of the equation
$\cos\theta + \cos 3\theta = \cos 2\theta$

11 Find the equation of the locus of the point P(x, y) when the distance of P from the line $y = 4$ is twice the distance between P and the point $(2, -1)$.

12 Find the Cartesian equation of the curve whose parametric equations are
$x = 2 - \cos 2\theta$ and $y = 5 + \sin\theta$

13 The equation of a circle is
$x^2 + y^2 + 2x - 6y - 3 = 0$
Find the equation of the diameter of this circle whose gradient is 1.

14 The focus of a parabola is the point $(2, 4)$ and the directrix is the line $y = 8$
Find the equation of the parabola.

15 (a) The equations of two circles are
$(x - 1)^2 + (y + 1)^2 = 9$
and $x^2 + y^2 + 4x - 6y + c = 0$
Find the value of c such that the circles touch.

   (b) Find the coordinates of the point of contact of the circles.

   (c) Find the equation of the common tangent through the point of contact.

16 (a) Prove that the line $3x - 2y - 12 = 0$ is a tangent to the circle
$(x - 1)^2 + (y - 2)^2 = 13$

   (b) Find the coordinates of the point of contact of the line and the circle.

17 Find the equations of the tangents from the origin to the circle $x^2 + y^2 - 8x - 6y + 16 = 0$

18 (a) Show that the line $y = 2x + 2$ is a tangent to the parabola whose parametric equations are
$x = 4t^2$, $y = 8t$

   (b) Find the coordinates of the point of contact of the line and parabola.

19 The equation of an ellipse is $4x^2 + 9y^2 - 36 = 0$

   (a) Sketch the ellipse.

   (b) Find the coordinates of the foci.

20 (a) Find the points of intersection of the line
$y = x + 5$ and the circle
$x^2 + y^2 - 2y - 25 = 0$

   (b) Find the equations of the tangents to the circle at each of these points.

21 The line l_1 has vector equation

$$\mathbf{r} = \begin{pmatrix} 1 \\ 2 \\ 4 \end{pmatrix} + \lambda \begin{pmatrix} 2 \\ 0 \\ 1 \end{pmatrix}$$

and the line l_2 has vector equation

$$\mathbf{r} = \begin{pmatrix} 2 \\ 1 \\ 1 \end{pmatrix} + \mu \begin{pmatrix} -3 \\ 1 \\ 2 \end{pmatrix}$$

where λ and μ are scalar parameters.

(a) Show that the lines intersect and give the position vector of their point of intersection.

(b) Find the acute angle between the two lines.

22 (a) The position vectors of three points are
$r = 2i - 3j + 2k$,
$r = 5i + k$,
$r = i + j - k$
Find the vector equation of the plane which contains these three points.

(b) Hence find the distance of the plane from the origin.

23 The equations of two lines are:
$x = 5 - \lambda, y = 2 + 3\lambda, z = -1 + \lambda$ and
$x = 4 + 2\mu, y = 3 - 6\mu, z = 1 - 2\mu$

(a) Show that the lines are parallel.

(b) Show that the vector $2i + j - k$ is perpendicular to these lines.

24 The Cartesian equation of a line l_1 is
$$\frac{x - 1}{2} = \frac{y - 2}{5} = \frac{z + 3}{2}$$
and the Cartesian equation of a line l_2 is
$$\frac{x + 1}{2} = \frac{y + 3}{3} = \frac{z - 1}{a}$$
Given that l_1 and l_2 are perpendicular, find the value of a.

25 The Cartesian equation of a plane is
$3x - y + 2z = 7$

(a) Find the distance of this plane from the origin.

(b) Show that the line whose vector equation is
$r = i + 2j + 4k + \lambda(2i + 2j - 3k)$
lies in the plane.
Hence write down the distance of the line from the origin.

26 The position vectors of points A and B are
$(2i - 2j + k)$ and $(3i + 6j - 2k)$ respectively relative to a fixed origin O.

Find the acute angle between OA and OB and hence find the area of triangle OAB.

27 Show that the planes whose equations are
$r \cdot (i - 5j + 2k) = 10$
and $r \cdot (-2i + 10j - 4k) = 8$ are parallel.
Hence find the distance between the two planes.

28 Show that the line whose vector equation is
$$r = \begin{pmatrix} 2 \\ 1 \\ 5 \end{pmatrix} + \lambda \begin{pmatrix} 1 \\ -1 \\ 4 \end{pmatrix}$$ is parallel to the plane
whose equation is $r \cdot \begin{pmatrix} 1 \\ 5 \\ 1 \end{pmatrix} = 8$

29 (a) Find the parametric equations of the line through the point $(2, 1, -5)$ that is perpendicular to the plane whose equation is
$3x + 3y - z = 9$

(b) Find the coordinates of the point of intersection of this line and the plane.

30 Show that if P(x, y) is twice as far from the point $(4, -2)$ as it is from the origin then P lies on a circle. Find the centre and radius of the circle.

31 P$(a \cos \theta_1, b \sin \theta_1)$ and Q$(a \cos \theta_2, b \sin \theta_2)$ are two points on the ellipse
$$\frac{x^2}{a^2} + \frac{y^2}{b^2} = 1$$

(a) Find the equation of the chord PQ.

(b) Deduce that the equation of the tangent to the ellipse at the point $(a \cos \theta, b \sin \theta)$ is
$ay \sin \theta + bx \cos \theta = ab$

32 P$(5t^2, 10t)$ is any point on a curve.

(a) Find the Cartesian equation of the curve.

(b) M is the midpoint of the line OP.
Find the Cartesian equation of the locus of M as t varies.

33 (a) Prove that
$$\cos^2 x + \sin^2 x \equiv 1$$

(b) Find the general solution of the equation
$$\cos^2 x + 5 \sin^2 x = 2$$

34 Show that the line whose vector equation is
$$r = \begin{pmatrix} 0 \\ 4 \\ 8 \end{pmatrix} + t \begin{pmatrix} 1 \\ 0 \\ -2 \end{pmatrix}$$
is contained in the plane whose
equation is $r \cdot \begin{pmatrix} 2 \\ 0 \\ 1 \end{pmatrix} = 8$

3.1 Functions – continuity and discontinuity

Learning outcomes

- To investigate the meaning of continuity and discontinuity of functions using graphs

You need to know

- The meaning of a function
- The meaning of the domain of a function
- The shapes of graphs of simple functions

Continuous functions

Most of the graphs that we have drawn so far in this unit have involved **continuous functions**. For a function to be continuous, there must be no breaks in its graph and no points at which it is undefined.

For example, $f(x) = x^2$ for $x \in \mathbb{R}$ is a continuous function because the graph of $y = f(x)$ has no breaks and $f(x)$ has a real value for every value of x in its domain.

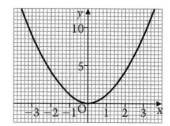

Discontinuity

The graph below illustrates the graph of a function over the domain $x \in \mathbb{R}$.

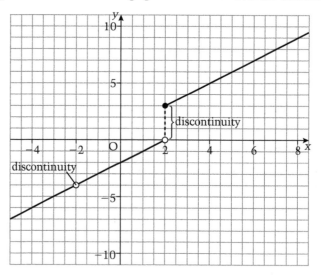

There is a clear break in the graph where $x = 2$. There is also a point missing from the graph where $x = -2$. These breaks are called **discontinuities**.

The function is continuous for all other values of x.

Some functions are continuous even though the graphs have breaks in them.

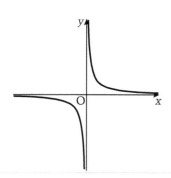

For example the function $f(x) = \dfrac{1}{x}$, $x \neq 0$, $x \in \mathbb{R}$ is continuous because although the graph of $y = \dfrac{1}{x}$ has a break in it where $x = 0$, this value of x is *not* in the domain of $f(x)$.

Example

(a) Sketch the graph of the function given by
$$f(x) = \begin{cases} x + 4, x \geqslant 4 \\ x - 4, x < 4 \end{cases}, x \in \mathbb{R}$$

(b) State, with a reason, whether the function is continuous over all of its domain.

(a)
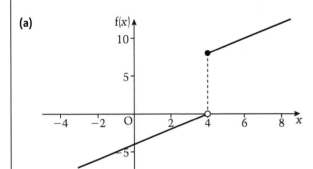

(b) There is a break where $x = 4$ and $x = 4$ is included in the domain, so the curve is not continuous over all the domain.

Example

(a) Sketch the graph of the function given by
$$f(x) = \begin{cases} x^2, x < 0 \\ x, \ x \geqslant 0 \end{cases}, x \in \mathbb{R}$$

(b) State whether the function is continuous.

(a)
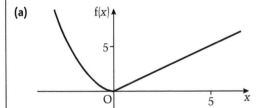

(b) There are no breaks in this graph and although there is a change in the nature of the graph where $x = 0$, there is not a point missing because $f(x)$ is defined when $x = 0$. Therefore the function is continuous.

Exercise 3.1

Sketch each of the following functions and state whether the function is continuous.

1 $f(x) = x, \ x = \{2, 4, 6\}$

2 $f(x) = \dfrac{1}{x + 1}, x \neq -1, x \in \mathbb{R}$

3 $f(x) = \begin{cases} x, \ \ \ \ \ x < 0 \\ x + 1, x \geqslant 0 \end{cases}, x \in \mathbb{R}$

4 $f(x) = \begin{cases} x + 1, \ \ \ x < 4 \\ -x + 9, x \geqslant 4 \end{cases}, x \in \mathbb{R}$

5 $f(x) = \begin{cases} x + 1, \ \ \ x < 4 \\ -x + 9, x \geqslant 5 \end{cases}, x \in \mathbb{R}$

Learning outcomes

- To introduce basic concepts of limits
- To investigate the behaviour of $f(x)$ as x approaches a from above and below

You need to know

- The meaning of continuity
- The shape of the curve $y = \dfrac{1}{x}$
- How to sketch simple functions

Limits

In Topic 1.10 we looked at the behaviour of $\dfrac{1}{x}$ as the value of x gets large. This table of values shows that $\dfrac{1}{x}$ approaches zero as x approaches infinity.

x	5	10	100	1000	10 000...
$\dfrac{1}{x}$	0.2	0.1	0.01	0.001	0.000 1...

We write this as $\dfrac{1}{x} \to 0$ as $x \to \infty$, and we say that the limiting value of $\dfrac{1}{x}$ as $x \to \infty$ is 0. The notation for this statement is $\displaystyle\lim_{x \to \infty} \dfrac{1}{x} = 0$

If we now look at $\dfrac{1}{x}$ as $x \to 0$, there are two cases to consider because x can approach zero from positive values or from negative values.

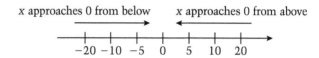

$$x \text{ approaches } 0 \text{ from below} \qquad x \text{ approaches } 0 \text{ from above}$$

As $x \to 0$ from above

$\left(\text{i.e. } \dfrac{1}{0.1} \, (= 10), \dfrac{1}{0.01} \, (= 100), \dfrac{1}{0.001} \, (= 1000)...\right), \dfrac{1}{x} \to \infty$

As $x \to 0$ from below

$\left(\text{i.e. } \dfrac{1}{-0.1} \, (= -10), \dfrac{1}{-0.01} \, (= -100), \dfrac{1}{-0.001} \, (= -1000)...\right), \dfrac{1}{x} \to -\infty$

Therefore $\displaystyle\lim_{x \to 0}\left(\dfrac{1}{x}\right)$ does not have a unique value so $\displaystyle\lim_{x \to 0}\left(\dfrac{1}{x}\right)$ does not exist.

> For $\displaystyle\lim_{x \to a}[f(x)]$ to exist and equal k then
>
> as $x \to a$ from above
> and as $x \to a$ from below $\left.\right\} f(x) \to k$

The limit of $f(x)$ as $x \to a$ from above is written as $\displaystyle\lim_{x \to a^+} f(x)$ and the limit of $f(x)$ as $x \to a$ from below is written as $\displaystyle\lim_{x \to a^-} f(x)$

Limits and discontinuity

We can now define a discontinuity in terms of limits.

For $f(x)$ to be continuous where $x = a$, $\displaystyle\lim_{x \to a^+} f(x) = \lim_{x \to a^-} f(x) = f(a)$

For example the function $f(x) = \begin{cases} x^2, & x < 0 \\ x, & x \geqslant 0 \end{cases}, x \in \mathbb{R}$,
is sketched in Topic 3.1.

Using the condition for continuity at $x = 0$,
$\lim\limits_{x \to 0^+} x = 0$, $\lim\limits_{x \to 0^-} x^2 = 0$ and

$f(0) = 0$, therefore $f(x)$ is continuous at $x = 0$

However applying this condition to the function

$f(x) = \begin{cases} x + 4, & x \geqslant 4 \\ x - 4, & x < 4 \end{cases}, x \in \mathbb{R}$, where $x = 4$, which is also sketched in

Topic 3.1, gives $\lim\limits_{x \to 4^+}(x + 4) = 8$ and $\lim\limits_{x \to 4^-}(x - 4) = 0$

They are not equal so there is a discontinuity at $x = 4$

Example

The function f is defined as $f(x) = \begin{cases} 2x - 1, & x > 1 \\ x^2, & x \leqslant 1 \end{cases}, x \in \mathbb{R}$.

(a) Find $\lim\limits_{x \to 1^+} f(x)$

(b) Find $\lim\limits_{x \to 1^-} f(x)$

(c) Hence show that $f(x)$ is continuous at $x = 1$

First sketch the graph $y = f(x)$

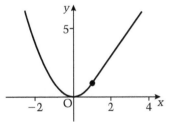

(a) From the graph, $\lim\limits_{x \to 1^+} f(x) = \lim\limits_{x \to 1^+}(2x - 1) = 1$

(b) From the graph, $\lim\limits_{x \to 1^-} f(x) = \lim\limits_{x \to 1^-} x^2 = 1$

(c) $f(1) = 1^2 = 1$

$\lim\limits_{x \to 1^+} f(x) = \lim\limits_{x \to 1^-} f(x) = f(1)$

Therefore $f(x)$ is continuous at $x = 1$

Exercise 3.2

(a) Sketch the graph of the function given by

$f(x) = \begin{cases} x^2, & x > 1 \\ 2 - x^2, & x \leqslant 1 \end{cases}, x \in \mathbb{R}$ for values of $-3 \leqslant x \leqslant 3$

(b) Find $\lim\limits_{x \to 1^+} f(x)$

(c) Find $\lim\limits_{x \to 1^-} f(x)$

(d) Hence show that $f(x)$ is continuous at $x = 1$

3.3 Limit theorems

Learning outcomes

- To list and use the limit theorems
- To find and use $\lim\limits_{\theta \to 0} \dfrac{\sin \theta}{\theta}$

You need to know

- The notation for limits
- The trigonometric double angle formulae and factor formulae
- How to factorise $x^3 - y^3$
- The factor theorem

The limit theorems

1 If $\lim\limits_{x \to a} f(x) = F$ then $\lim\limits_{x \to a} kf(x) = kF$ where k is a constant.

For example, $\lim\limits_{x \to 2} x^2 = 4$, therefore $\lim\limits_{x \to 2} 3x^2 = 3 \times 4 = 12$

2 If $\lim\limits_{x \to a} f(x) = F$ and if $\lim\limits_{x \to a} g(x) = G$ then $\lim\limits_{x \to a}[f(x) \times g(x)] = FG$

For example, $\lim\limits_{x \to 2} x^2 = 4$ and $\lim\limits_{x \to 2} (x + 1) = 3$,

therefore $\lim\limits_{x \to 2} [x^2(x + 1)] = 4 \times 3 = 12$

3 If $\lim\limits_{x \to a} f(x) = F$ and if $\lim\limits_{x \to a} g(x) = G$ then $\lim\limits_{x \to a}[f(x) + g(x)] = F + G$

For example, $\lim\limits_{x \to 2} x^2 = 4$ and $\lim\limits_{x \to 2} (x + 1) = 3$,

therefore $\lim\limits_{x \to 2} [x^2 + x + 1] = 4 + 3 = 7$

4 If $\lim\limits_{x \to a} f(x) = F$ and if $\lim\limits_{x \to a} g(x) = G$ then, provided $G \neq 0$, $\lim\limits_{x \to a}\left(\dfrac{f(x)}{g(x)}\right) = \dfrac{F}{G}$

For example, $\lim\limits_{x \to 2} x^2 = 4$ and $\lim\limits_{x \to 2} (x + 1) = 3$,

therefore $\lim\limits_{x \to 2} \dfrac{x^2}{x + 1} = \dfrac{4}{3}$

Now $\dfrac{0}{0}$ is meaningless, but these theorems can be used to find the limit of a function which appears to be $\dfrac{0}{0}$. The example below illustrates this.

Example

Find $\lim\limits_{x \to 3} \left(\dfrac{x^2 - 9}{x^2 - 7x + 12}\right)$

$\dfrac{x^2 - 9}{x^2 - 7x + 12} = \dfrac{(x - 3)(x + 3)}{(x - 3)(x - 4)} = \dfrac{x + 3}{x - 4}$ provided that $x \neq 3$

The limit as x approaches 3 means we want the values of the function as x gets closer and closer to 3 to see what value they are tending towards. We do not want the value when $x = 3$, so we can say that

$$\lim\limits_{x \to 3} \left(\dfrac{x^2 - 9}{x^2 - 7x + 12}\right) = \lim\limits_{x \to 3} \left(\dfrac{x + 3}{x - 4}\right)$$

$$= \dfrac{\lim\limits_{x \to 3} (x + 3)}{\lim\limits_{x \to 3} (x - 4)} = \dfrac{6}{-1} = -6$$

The limit of $\dfrac{\sin x}{x}$ as $x \to 0$

This is an important limit and the limit theorems do not help to find it directly, so we start with the graphs of $y = \sin x$ and $y = x$ close to the origin.

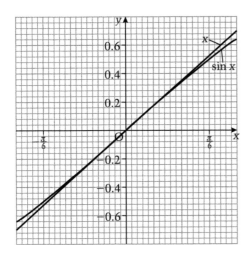

When $x = 0$, $\dfrac{\sin x}{x} = \dfrac{0}{0}$ and $\dfrac{0}{0}$ is meaningless.

For $-\dfrac{\pi}{6} < x < \dfrac{\pi}{6}$, $\sin x$ and x are nearly equal, and as x approaches 0 from above, $\dfrac{\sin x}{x} \approx 1$

Also as x approaches 0 from below, $\dfrac{\sin x}{x} \approx 1$

$$\textbf{Therefore } \lim_{x \to 0} \frac{\sin x}{x} = 1$$

Note that this result is valid *only* when x is measured in radians.

Example

Find $\displaystyle\lim_{\theta \to 0} \dfrac{\sin 2\theta}{\theta}$

$$\frac{\sin 2\theta}{\theta} = \frac{2 \sin \theta \cos \theta}{\theta} = \frac{\sin \theta}{\theta} \times 2\cos\theta$$

$$\therefore \quad \lim_{\theta \to 0} \frac{\sin 2\theta}{\theta} = \lim_{\theta \to 0} \frac{\sin \theta}{\theta} \times \lim_{\theta \to 0} (2\cos\theta) = 1 \times 2 = 2$$

Alternatively, $\displaystyle\lim_{\theta \to 0} \frac{\sin 2\theta}{\theta} = \lim_{\theta \to 0} \frac{2 \sin 2\theta}{2\theta} = 2\lim_{\theta \to 0} \frac{\sin 2\theta}{2\theta} = 2$

Exercise 3.3

1 Find the limit of $\dfrac{x^2 - 5x + 6}{x - 2}$ as $x \to 2$

2 Find the limit of $\dfrac{x^2 - 1}{x^3 - x^2 + x - 1}$ as $x \to 1$

3 Find the limit of $\dfrac{\sin 3\theta - \sin \theta}{2\theta}$ as $\theta \to 0$

4 Use the results above to show that the function $f(x) = \dfrac{\sin x}{x}$ has a discontinuity where $x = 0$

Learning outcomes

- Find the gradient of a curve at a point on the curve

You need to know

- The definition of the gradient of a straight line
- The concept of a limit

Tangents, chords and normals

The line joining two points on a curve is a *chord*.

A line that touches a curve at a point is a *tangent* to the curve.

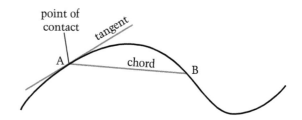

The gradient of a curve at a given point

The gradient of a curve at a point A is defined as the gradient of the tangent to the curve at A.

We can find the gradient of a curve at a point A by taking another point B on the curve close to A.

As B moves closer to A, the gradient of the chord gets closer to the gradient of the tangent at A.

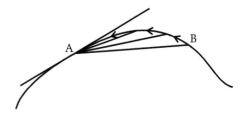

$$\underset{B \to A}{\text{limit}}\ (\text{gradient of chord AB}) = \text{gradient of tangent at A}$$

We can use this to find the gradient of the curve $y = x^2$ at the point where $x = \frac{1}{2}$.

A is the point on the curve where $x = \frac{1}{2}$,

B is a point on the curve whose x-coordinate is a bit larger than $\frac{1}{2}$.

We denote the 'bit larger' by δx.   *h* is sometimes used as an alternative notation for the 'bit larger'

So B is the point on the curve where $x = \frac{1}{2} + \delta x$

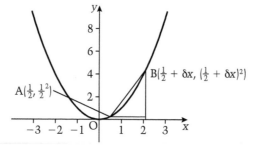

δ is not a variable – it is a prefix that means 'a small increase in' the value of the variable that follows it. So δy means a small increase in the value of y, δt means a small increase in the value of t, and so on.

The coordinates of A are $(\frac{1}{2}, \frac{1}{4})$

The coordinates of B are $(\frac{1}{2} + \delta x, (\frac{1}{2} + \delta x)^2)$

The gradient of AB is $\dfrac{(\frac{1}{2} + \delta x)^2 - \frac{1}{4}}{\frac{1}{2} + \delta x - \frac{1}{2}}$

$$= \dfrac{\frac{1}{4} + \delta x + (\delta x)^2 - \frac{1}{4}}{\delta x} = 1 + \delta x$$

$\therefore$ gradient of tangent at A $= \underset{B \to A}{\text{limit}}$ (gradient of chord AB)

$$= \underset{\delta x \to 0}{\text{limit}} (1 + \delta x) = 1$$

$\therefore$ at the point where $x = \frac{1}{2}$, the gradient of $y = x^2$ is 1.

We can apply the same process to a variable point on $y = x^2$

A is any point on the curve so its coordinates can be denoted by (x, x^2).

B is the point on the curve whose x-coordinate is a little larger than the x-coordinate of A,

i.e. $x + \delta x$. Hence B has coordinates $(x + \delta x, (x + \delta x)^2)$

The gradient of AB is $\dfrac{(x + \delta x)^2 - x^2}{(x + \delta x) - x}$

$$= \dfrac{x^2 + 2x\delta x + (\delta x)^2 - x^2}{\delta x}$$

$$= \dfrac{2x\delta x + (\delta x)^2}{\delta x} = \dfrac{\delta x(2x + \delta x)}{\delta x}$$

$$= 2x + \delta x$$

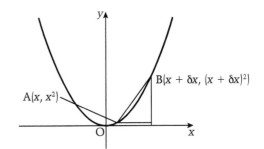

The gradient of the tangent at A $= \lim$ (gradient of AB)

$$= \underset{\delta x \to 0}{\lim} (2x + \delta x) = 2x$$

We can now use this result to find the gradient at any particular point on $y = x^2$

For example, at the point where $x = 3$, the gradient of the curve is $2 \times 3 = 6$, and where $x = -4$ the gradient is $2 \times -4 = -8$

Exercise 3.4

1 Use the method above to find the gradient at any point on the curve $y = x^3$

   Use your result to find the gradient on the curve where:

   (a) $x = -1$ **(b)** $x = 5$

2 Use the method above to find the gradient at any point on the curve $y = x^2 + 2x$

   Use your result to find the gradient at the points where:

   (a) $x = 0$ **(b)** $x = -\frac{1}{2}$

The gradient function

In the previous topic, we found that the gradient of any point on the curve $y = x^2$ is given by $2x$.

Now $2x$ is a function and it is called the gradient function of x^2.

Because $2x$ is derived from x^2, $2x$ is often called the **derived function** or the **derivative**.

The gradient function of a general curve $y = f(x)$ is denoted by $f'(x)$ or by $\dfrac{dy}{dx}$.

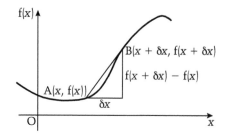

So for $y = x^2$, we write $f'(x) = 2x$ or $\dfrac{dy}{dx} = 2x$

For any curve $y = f(x)$, gradient of AB $= \dfrac{f(x + \delta x) - f(x)}{\delta x}$

and the gradient at A $= \displaystyle\lim_{\delta x \to 0} \dfrac{f(x + \delta x) - f(x)}{\delta x}$

i.e.
$$f'(x) = \lim_{\delta x \to 0} \frac{f(x + \delta x) - f(x)}{\delta x}$$

We can use this general formula to find the gradient of any function, and it is called **differentiation from first principles**.

Example

Differentiate $\dfrac{1}{x^2}$ from first principles.

$f(x) = \dfrac{1}{x^2}$, so $f(x + \delta x) = \dfrac{1}{(x + \delta x)^2}$

$f(x + \delta x) - f(x) = \dfrac{1}{(x + \delta x)^2} - \dfrac{1}{x^2} = \dfrac{x^2 - (x + \delta x)^2}{x^2(x + \delta x)^2} = \dfrac{-2x\delta x - (\delta x)^2}{x^2(x + \delta x)^2}$

$\therefore \quad f'(x) = \displaystyle\lim_{\delta x \to 0} \dfrac{f(x + \delta x) - f(x)}{\delta x} = \lim_{\delta x \to 0} \dfrac{-2x\delta x - (\delta x)^2}{\delta x(x + \delta x)^2 x^2}$

$\qquad\qquad = \displaystyle\lim_{\delta x \to 0} \dfrac{-2x - \delta x}{x^2(x + \delta x)^2} = \dfrac{-2x}{x^4} = -\dfrac{2}{x^3}$

$\therefore \quad$ when $y = \dfrac{1}{x^2}$, $\dfrac{dy}{dx} = -\dfrac{2}{x^3}$

Example

Differentiate $\sin x$ from first principles.

$f(x) = \sin x$, so $f(x + \delta x) = \sin(x + \delta x)$

$f(x + \delta x) - f(x) = \sin(x + \delta x) - \sin x$ Using a factor formula

$\qquad\qquad = 2\cos\left(x + \tfrac{1}{2}\delta x\right)\sin\tfrac{1}{2}\delta x$

$\therefore\ f'(x) = \lim_{\delta x \to 0}\dfrac{2\cos\left(x + \tfrac{1}{2}\delta x\right)\sin\tfrac{1}{2}\delta x}{\delta x} = \lim_{\delta x \to 0}\dfrac{\cos\left(x + \tfrac{1}{2}\delta x\right)\sin\tfrac{1}{2}\delta x}{\tfrac{1}{2}\delta x}$

Now as $\delta x \to 0$, $\cos\left(x + \tfrac{1}{2}\delta x\right) \to \cos x$ and, provided that x is in

radians, $\dfrac{\sin\tfrac{1}{2}\delta x}{\tfrac{1}{2}\delta x} \to 1$

$\therefore\ f'(x) = \cos x$

i.e. if $y = \sin x$, then $\dfrac{dy}{dx} = \cos x$

Example

Differentiate $\sqrt{x}$ from first principles and hence find the gradient of the curve at the point where $x = 4$

$f(x) = \sqrt{x} = x^{\frac{1}{2}}$

$f(x + \delta x) - f(x) = (x + \delta x)^{\frac{1}{2}} - x^{\frac{1}{2}}$

$\qquad = \dfrac{\left((x + \delta x)^{\frac{1}{2}} - x^{\frac{1}{2}}\right)\left((x + \delta x)^{\frac{1}{2}} + x^{\frac{1}{2}}\right)}{\left((x + \delta x)^{\frac{1}{2}} + x^{\frac{1}{2}}\right)}$

$\qquad = \dfrac{x + \delta x - x}{\left((x + \delta x)^{\frac{1}{2}} + x^{\frac{1}{2}}\right)}$ Using $(a + b)(a - b) = a^2 - b^2$

$\qquad = \dfrac{\delta x}{\left((x + \delta x)^{\frac{1}{2}} + x^{\frac{1}{2}}\right)}$

$\therefore\ f'(x) = \lim_{\delta x \to 0}\dfrac{\delta x}{\delta x\left((x + \delta x)^{\frac{1}{2}} + x^{\frac{1}{2}}\right)} = \lim_{\delta x \to 0}\dfrac{1}{(x + \delta x)^{\frac{1}{2}} + x^{\frac{1}{2}}} = \dfrac{1}{2x^{\frac{1}{2}}} = \dfrac{1}{2}x^{-\frac{1}{2}}$

i.e. if $y = x^{\frac{1}{2}}$, $\dfrac{dy}{dx} = \dfrac{1}{2}x^{-\frac{1}{2}}$

When $x = 4$, $\dfrac{dy}{dx} = \dfrac{1}{2}(4)^{-\frac{1}{2}} = \dfrac{1}{4}$

Exercise 3.5

Differentiate the following functions from first principles.

1 $f(x) = kx$ where $k \in \mathbb{R}$

2 $f(x) = \cos x$

3 $f(x) = 5x^3$

4 $f(x) = x^{-1}$

5 $(x^2 - 2x + 1)$

6 $f(x) = \sin 2x$

- To derive rules for differentiating simple functions

- The equations of straight lines
- The meaning of $\dfrac{dy}{dx}$

Differentiation of a constant

The equation $y = c$, where c is a constant, represents a straight line parallel to the x-axis.

Therefore the gradient of the line is zero,

i.e.   **when $y = c$, $\dfrac{dy}{dx} = 0$**

For example, when $y = -5$, $\dfrac{dy}{dx} = 0$

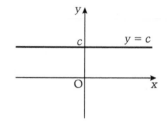

Differentiation of $y = ax$

The equation $y = ax$, where a is a constant, represents a line through the origin with gradient a.

Therefore when $y = ax$, $\dfrac{dy}{dx} = a$

For example, when $y = 4x$, $\dfrac{dy}{dx} = 4$

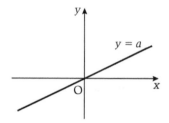

Differentiation of $y = x^n$

The table shows some of the results from Topic 3.5:

$y = f(x)$	x^3	x^2	$x^{\frac{1}{2}}$	x^{-1}	x^{-2}
$\dfrac{dy}{dx}$	$3x^2$	$2x$	$\dfrac{1}{2}x^{-\frac{1}{2}}$	$-x^{-2}$	$-2x^{-3}$

These results suggest that to differentiate a power of x, we multiply by that power and reduce the power by 1,

i.e.   **when $y = x^n$, $\dfrac{dy}{dx} = nx^{n-1}$ for all values of n.**

For example, when $y = x^{10}$, $\dfrac{dy}{dx} = 10x^9$ and when $y = x^{-4}$, $\dfrac{dy}{dx} = -4x^{-5}$

Differentiation of $y = ax^n$

The result from Exercise 3.5, question 3, shows that when $y = 5x^3$,
$\dfrac{dy}{dx} = 15x^2 = 5 \times 3x^2$

This is a particular example of the general result,

i.e.   **when $y = ax^n$, $\dfrac{dy}{dx} = anx^{n-1}$ where a is a constant.**

For example, when $y = 4x^{-\frac{1}{2}}$, $\dfrac{dy}{dx} = 4 \times -\dfrac{1}{2}x^{-\frac{1}{2}-1} = -2x^{-\frac{3}{2}}$

Differentiation of $y = f(x) + g(x)$

The result from Exercise 3.5, question 5, shows that when

$y = x^2 - 2x + 1, \dfrac{dy}{dx} = 2x - 2$

This is the same as differentiating each term separately, i.e.

$\dfrac{dy}{dx} = \dfrac{d}{dx}(x^2) - \dfrac{d}{dx}(2x) + \dfrac{d}{dx}(1)$

The notation $\dfrac{d}{dx}f(x)$ means the differential of $f(x)$ with respect to x, i.e. $\dfrac{d}{dx}f(x) = f'(x)$

This result is true for the differential of the sum or difference of any functions,

i.e **when $y = f(x) \pm g(x), \dfrac{dy}{dx} = \dfrac{d}{dx}f(x) \pm \dfrac{d}{dx}g(x)$**

Two other results from Topic 3.5 are important:

when $y = \sin x, \dfrac{dy}{dx} = \cos x$ and   when $y = \cos x, \dfrac{dy}{dx} = -\sin x$

All these results are important and you need to remember them.

We have used the letters y and x for the variables, but any letters can be used, for example when $s = 2t + \cos t, \dfrac{ds}{dt} = 2 - \sin t$

(Letters s and t are often used for displacement and time in problems concerned with movement.)

Example

Find $\dfrac{dy}{dx}$ when $y = \dfrac{6x^3 - 5x + 2}{3x}$

$\dfrac{6x^3 - 5x + 2}{3x} = \dfrac{6x^3}{3x} - \dfrac{5x}{3x} + \dfrac{2}{3x}$

$\qquad\qquad = 2x^2 - \dfrac{5}{3} + \dfrac{2}{3}x^{-1}$

$\therefore y = 2x^2 - \dfrac{5}{3} + \dfrac{2}{3}x^{-1}$ so $\dfrac{dy}{dx} = 4x - \dfrac{2}{3}x^{-2}$

✅ *Exam tip*

$\dfrac{d}{dx}\left(\dfrac{f(x)}{g(x)}\right) \neq \dfrac{\dfrac{d}{dx}f(x)}{\dfrac{d}{dx}g(x)}$

Exercise 3.6

Find $\dfrac{dy}{dx}$ when y is:

1   $5x^4$

2   $\dfrac{1}{x^2}$

3   $4x^3 - 2x + 5$

4   $(3x - 4)(2x + 1)$

5   $4 - 3\cos x$

6   $\dfrac{2x^3 - 3}{x}$

7   $x(2 - \sqrt{x})$

8   $5\sin x - 4\cos x$

✅ *Exam tip*

$\dfrac{d}{dx}f(x) \times g(x) \neq \dfrac{d}{dx}f(x) \times \dfrac{d}{dx}g(x)$

- To derive and use a formula for differentiating a product of functions
- To derive and use a formula for differentiating a quotient of functions

You need to know

- How to differentiate sums and differences of powers of x
- How to differentiate $\sin x$ and $\cos x$
- The limit theorems
- Trigonometric identities

The rule for differentiating a product of functions

If $y = uv$, where $u = f(x)$ and $v = g(x)$

and if δx is a small increase in the value of x, and δy, δu and δv are the corresponding increases in y, u and v, then

$$y + \delta y = (u + \delta u)(v + \delta v)$$

$$= uv + u(\delta v) + v(\delta u) + (\delta u)(\delta v)$$

As $y = uv$, this simplifies to

$$\delta y = u(\delta v) + v(\delta u) + (\delta u)(\delta v)$$

Dividing by δx gives

$$\frac{\delta y}{\delta x} = u\frac{\delta v}{\delta x} + v\frac{\delta u}{\delta x} + \delta u\frac{\delta v}{\delta x}$$

Now $\lim\limits_{\delta x \to 0} \dfrac{\delta y}{\delta x} = \dfrac{dy}{dx}$, $\lim\limits_{\delta x \to 0} \dfrac{\delta v}{\delta x} = \dfrac{dv}{dx}$, $\lim\limits_{\delta x \to 0} \dfrac{\delta u}{\delta x} = \dfrac{du}{dx}$ and $\lim\limits_{x \to 0} \delta u = 0$

Therefore $\dfrac{dy}{dx} = \lim\limits_{x \to 0}\left(\dfrac{\delta y}{\delta x}\right) = u\dfrac{dv}{dx} + v\dfrac{du}{dx}$,

i.e.
$$\frac{d}{dx}(uv) = u\frac{dv}{dx} + v\frac{du}{dx}$$

You may find it easier to remember this rule in words:

> To differentiate a product of functions,
> multiply the first function by the differential of the second function
> then add the second function multiplied by the differential of the first.

For example, if $y = x^2 \cos x$, then using $u = x^2$ and $v = \cos x$ gives

$$\frac{dy}{dx} = -x^2 \sin x + (\cos x)(2x) = 2x \cos x - x^2 \sin x$$

Exercise 3.7a

Use the product rule to find $\dfrac{dy}{dx}$ when y is:

(a) $x^3 \sin x$

(b) $\sin x \cos x$

(c) $\dfrac{1}{x} \sin x$

(d) $x^{\frac{1}{2}}(3x^3 - 4)$

(e) $(x^2 + 1)\cos x$

The rule for differentiating a quotient of functions

If $y = \dfrac{u}{v}$, where $u = f(x)$ and $v = g(x)$

and if δx is a small increase in the value of x, and δy, δu and δv are the corresponding increases in y, u and v,

then $y + \delta y = \dfrac{u + \delta u}{v + \delta v}$

$$y = \frac{u}{v}, \quad \therefore \quad \delta y = \frac{u + \delta u}{v + \delta v} - \frac{u}{v}$$

$$= \frac{v\delta u - u\delta v}{v^2 + v\delta v}$$

Dividing by δx gives

$$\frac{\delta y}{\delta x} = \frac{v\dfrac{\delta u}{\delta x} - u\dfrac{\delta v}{\delta x}}{v^2 + v\delta v}$$

Now $\lim\limits_{\delta x \to 0} \dfrac{\delta y}{\delta x} = \dfrac{dy}{dx}$, $\lim\limits_{\delta x \to 0} \dfrac{\delta v}{\delta x} = \dfrac{dv}{dx}$, $\lim\limits_{\delta x \to 0} \dfrac{\delta u}{\delta x} = \dfrac{du}{dx}$ and $\lim\limits_{\delta x \to 0} \delta v = 0$

Therefore $\dfrac{dy}{dx} = \lim\limits_{\delta x \to 0}\left(\dfrac{\delta y}{\delta x}\right) = \dfrac{v\dfrac{du}{dx} - u\dfrac{dv}{dx}}{v^2}$

i.e. **if $y = \dfrac{u}{v}$ then $\dfrac{dy}{dx} = \dfrac{v\dfrac{du}{dx} - u\dfrac{dv}{dx}}{v^2}$**

You need to remember this rule. The order in the numerator is important. One way to remember it is by using the old word for a computer monitor – 'visual display unit' or VDU. So, VDU comes first.

For example, if $y = \dfrac{\sin x}{x^2}$, then using $u = \sin x$ and $v = x^2$ gives

$$\frac{dy}{dx} = \frac{x^2 \cos x - 2x \sin x}{x^4} = \frac{x \cos x - 2 \sin x}{x^3}$$

Alternatively, writing y as a quotient, i.e. $y = x^{-2} \sin x$, and using the

product rule gives $\dfrac{dy}{dx} = -2x^{-3} \sin x + x^{-2} \cos x = \dfrac{x \cos x - 2 \sin x}{x^3}$

The disadvantage of writing a quotient as a product is that the simplification of the result is often complicated.

Exercise 3.7b

Use the quotient rule to find $\dfrac{dy}{dx}$ when y is:

(a) $\tan x \left(= \dfrac{\sin x}{\cos x}\right)$

(b) $\dfrac{x^2}{x - 1}$

(c) $\dfrac{\cos x}{\sqrt{x}}$

(d) $\dfrac{x}{x^2 + 1}$

(e) $\cot x$

Learning outcomes

■ To derive the chain rule and to use it to differentiate composite functions

You need to know

■ The limit theorems

■ The meaning of a composite function

■ The differentials of simple functions

■ How to differentiate a product and quotient of functions

Differentiating a composite function

When y is a composite function, i.e. when $y = \text{gf}(x)$ where $u = \text{f}(x)$, we can write $y = g(u)$.

If δx is a small increase in the value of x and δy and δu are the corresponding increases in y and u, then

$$\frac{\delta y}{\delta x} = \frac{\delta y}{\delta u} \times \frac{\delta u}{\delta x}$$

As $\delta x \to 0, \delta y$ and δu also approach zero.

Therefore
$$\frac{dy}{dx} = \lim_{\delta x \to 0}\left(\frac{\delta y}{\delta x}\right) = \lim_{\delta x \to 0}\left(\frac{\delta y}{\delta u} \times \frac{\delta u}{\delta x}\right)$$

$$= \lim_{\delta u \to 0}\left(\frac{\delta y}{\delta u}\right) \times \lim_{\delta u \to 0}\left(\frac{\delta u}{\delta x}\right)$$

$$= \frac{dy}{du} \times \frac{du}{dx}$$

i.e
$$\boldsymbol{\frac{dy}{dx} = \frac{dy}{du} \times \frac{du}{dx}}$$

This is known as the ***chain rule*** and can be used to differentiate a composite function $\text{gf}(x)$ by making the substitution $u = \text{f}(x)$

For example, when $y = (2x + 3)^6$, using $u = 2x + 1$ gives $y = u^6$

then $\dfrac{dy}{du} = 6u^5$ and $\dfrac{du}{dx} = 2$

$\dfrac{dy}{dx} = \dfrac{dy}{du} \times \dfrac{du}{dx}$ gives $\dfrac{dy}{dx} = 6u^5 \times 2 = 12u^5$

Substituting $2x + 1$ for u gives $\dfrac{dy}{dx} = 12(2x + 1)^5$

You will not usually be given the substitution, so you need to recognise $\text{f}(x)$ when $y = \text{gf}(x)$

Example

Find $\text{f}'(x)$ when $\text{f}(x) = \sqrt{2x^2 - 5}$

Let $y = \sqrt{2x^2 - 5}$ then if $u = 2x^2 - 5$, $y = \sqrt{u} = u^{\frac{1}{2}}$

Therefore $\dfrac{du}{dx} = 4x$ and $\dfrac{dy}{du} = \dfrac{1}{2}u^{-\frac{1}{2}} = \dfrac{1}{2\sqrt{u}}$

$\dfrac{dy}{dx} = \dfrac{dy}{du} \times \dfrac{du}{dx}$ gives $\dfrac{dy}{dx} = \dfrac{1}{2\sqrt{u}} \times 4x$

$$= \frac{2x}{\sqrt{u}}$$

$$\therefore \quad \text{f}'(x) = \frac{2x}{\sqrt{2x^2 - 5}}$$

Example

Differentiate $\sin\left(3\theta + \frac{\pi}{4}\right)$ with respect to θ.

When θ is the variable, we replace x with θ.

Let $y = \sin\left(3\theta + \frac{\pi}{4}\right)$ and $u = 3\theta + \frac{\pi}{4}$ then $y = \sin u$

Using $\dfrac{dy}{d\theta} = \dfrac{du}{d\theta} \times \dfrac{d\theta}{du}$ with $\dfrac{du}{d\theta} = 3$ and $\dfrac{dy}{du} = \cos u$ gives

$$\frac{dy}{d\theta} = (\cos u)(3)$$
$$= 3\cos\left(3\theta + \frac{\pi}{4}\right)$$

After a bit of practice with simpler functions, you should be able to make the substitution mentally and write down the differential directly.

For example, you could go straight from

$y = 2(3x - 1)^3$ to $\dfrac{dy}{dx} = (2)(3)(3)(3x - 1)^2$
$$= 18(3x - 1)^2$$

Example

Find $\dfrac{dy}{dx}$ when $y = \dfrac{2}{(x - 3)}$

We could use the quotient rule for this, but by writing the equation as $y = 2(x - 3)^{-1}$, we can use the chain rule: i.e.

$y = 2(x - 3)^{-1} \Rightarrow \dfrac{dy}{dx} = (2)(-1)(x - 3)^{-2}$

$\Rightarrow \dfrac{dy}{dx} = -\dfrac{2}{(x - 3)^2}$

Exercise 3.8a

1 Differentiate each function with respect to x.

 (a) $(3x + 4)^4$

 (b) $(2 - x)^3$

 (c) $\sin 2x$

 (d) $\cos^3 x$

 (e) $\dfrac{1}{\sin x}$

2 Differentiate $\cos^4 \theta$ with respect to θ.

3 Find $\dfrac{dy}{dx}$ when $y = \dfrac{1}{\sqrt{x^2 + 1}}$

To differentiate more complicated functions, it is sensible to write down the substitutions.

Example

Find $\dfrac{dy}{dx}$ given $y = (3x - 2)^2 \sin\left(2x - \dfrac{\pi}{3}\right)$

This is the product of two composite functions. We will start by finding the differential of each composite function and then apply the product rule.

Let $t = (3x - 2)^2$ and $u = 3x - 2$ so $t = u^2$

$\Rightarrow \quad \dfrac{dt}{du} = 2u$ and $\dfrac{du}{dx} = 3$

then $\dfrac{dt}{dx} = \dfrac{dt}{du} \times \dfrac{du}{dx}$

$\qquad = 6u$

$\qquad = 6(3x - 2)$

Let $s = \sin\left(2x - \dfrac{\pi}{3}\right)$ and $u = \left(2x - \dfrac{\pi}{3}\right)$ so $s = \sin u$

$\Rightarrow \quad \dfrac{du}{dx} = 2$ and $\dfrac{ds}{du} = \cos u$

then $\dfrac{ds}{dx} = \dfrac{ds}{du} \times \dfrac{du}{dx}$

$\qquad = (\cos u)(2)$

$\qquad = 2 \cos\left(2x - \dfrac{\pi}{3}\right)$

Now using the product rule gives

$\dfrac{dy}{dx} = t\dfrac{ds}{dx} + s\dfrac{dt}{dx}$

$\qquad = (3x - 2)^2 \times 2 \cos\left(2x - \dfrac{\pi}{3}\right) + \sin\left(2x - \dfrac{\pi}{3}\right) \times 6(3x - 2)$

$\qquad = 2(3x - 2)\left[(3x - 2) \cos\left(2x - \dfrac{\pi}{3}\right) + 3 \sin\left(2x - \dfrac{\pi}{3}\right)\right]$

Extending the chain rule

We can extend the chain rule to cover functions that are a composite of three functions, i.e. where $y = \text{hgf}(x)$, by using $y = h(v)$ and $v = g(u)$ where $u = f(x)$ and then using the extended version of the chain rule, i.e.

$$\dfrac{dy}{dx} = \dfrac{dy}{dv} \times \dfrac{dv}{du} \times \dfrac{du}{dx}$$

For example, when $y = \sqrt{\cos(x^2 + 1)}$,

then $u = x^2 + 1$ and $v = \cos u$ so $y = v^{\frac{1}{2}}$

then $\quad \dfrac{dy}{dx} = \frac{1}{2}v^{-\frac{1}{2}} \times (-\sin u) \times 2x$

$\qquad\qquad = \frac{1}{2}\cos^{-\frac{1}{2}}u \times (-\sin u) \times 2x$

$\qquad\qquad = -\dfrac{x\sin(x^2 + 1)}{\sqrt{\cos(x^2 + 1)}}$

Example

Differentiate $\dfrac{1}{\sin^2(3x - 1)}$

First we can write $\dfrac{1}{\sin^2(3x - 1)}$ as $\sin^{-2}(3x - 1)$

Now $\sin^{-2}(3x - 1) = \mathrm{hgf}(x)$

where $f(x) = (3x - 1)$,

$g(x) = \sin x$ and $h(x) = x^{-2}$

so we need two substitutions.

Let $y = \sin^{-2}(3x - 1)$,

$u = 3x - 1$

and $v = \sin u$

then $y = v^{-2}$

then $\dfrac{dy}{dx} = \dfrac{dy}{dv} \times \dfrac{dv}{du} \times \dfrac{du}{dx}$

gives $\dfrac{dy}{dx} = -2v^{-3} \times \cos u \times 3$

$\qquad\qquad = -6(\sin u)^{-3}\cos u$

$\qquad\qquad = -\dfrac{6\cos(3x - 1)}{\sin^3(3x - 1)}$

Exercise 3.8b

1 Differentiate $(2x - 1)^4\sqrt{x^2 - 1}$ with respect to x.

2 Find $\dfrac{dy}{d\theta}$ when $y = \cos\left(2\theta - \frac{\pi}{4}\right)\sin^2\theta$

3 Find $f'(x)$ when $f(x) = \sin\left[(x + 4)^5\right]$

4 Find $\dfrac{dy}{dx}$ when $y = \sqrt{1 + (3x - 1)^4}$

Learning outcomes

- To find $\dfrac{dy}{dx}$ when the equation of a curve is given parametrically

You need to know

- The meaning of parametric equations
- How to differentiate simple functions
- The product rule, quotient rule and chain rule
- The limit theorems

The formula $\dfrac{dy}{dx} = 1 \div \dfrac{dx}{dy}$

This is a formula we need to be able to differentiate a *parametric equation*.

When $y = f(x)$, then $\quad \dfrac{dy}{dx} = \lim\limits_{\delta x \to 0}\dfrac{\delta y}{\delta x} = \lim\limits_{\delta x \to 0}\left(1 \div \dfrac{\delta x}{\delta y}\right)$

But $\delta y \to 0$ as $\delta x \to 0$ so $\quad \dfrac{dy}{dx} = 1 \div \lim\limits_{\delta y \to 0}\left(\dfrac{\delta x}{\delta y}\right) = 1 \div \dfrac{dx}{dy}$

i.e.
$$\dfrac{dy}{dx} = 1 \div \dfrac{dx}{dy}$$

Parametric differentiation

When the equation of a curve is given parametrically, we can find $\dfrac{dy}{dx}$ in terms of the parameter.

If $x = f(t)$ and $y = g(t)$, then the chain rule gives $\dfrac{dy}{dx} = \dfrac{dy}{dt} \times \dfrac{dt}{dx}$

Using the formula above, $\dfrac{dt}{dx} = 1 \div \dfrac{dx}{dt}$

so $\dfrac{dy}{dx} = \dfrac{dy}{dt} \times \dfrac{dt}{dx}$ becomes $\dfrac{dy}{dx} = \dfrac{dy}{dt} \div \dfrac{dx}{dt}$,

then $\dfrac{dy}{dx} = \dfrac{g'(t)}{f'(t)}$

For example, when $x = t^2$ and $y = \dfrac{1}{t+1}$ then

$\dfrac{dx}{dt} = 2t$ and $\dfrac{dy}{dt} = -\dfrac{1}{(t+1)^2}$ Using the quotient rule

$\therefore \quad \dfrac{dy}{dx} = \dfrac{dy}{dt} \div \dfrac{dx}{dt}$

$\qquad = -\dfrac{1}{(t+1)^2} \div 2t$

$\qquad = -\dfrac{1}{2t(t+1)^2}$

Exercise 3.9a

Find $\dfrac{dy}{dx}$ in terms of the parameter when:

(a) $y = t^2, x = 1 - t^3$

(b) $y = \cos\theta, x = 2\sin\theta$

(c) $x = \dfrac{1}{t}, y = (1+t)^2$

General differentiation

In the previous topics we have given the differentials of various types of functions and derived rules for combinations of functions. These results are summarised in the table.

y	$\dfrac{dy}{dx}$	General results
c (a constant)	0	$\dfrac{d}{dx}(f(x)+g(x))=f'(x)+g'(x)$
x^n	nx^{n-1}	$\dfrac{d}{dx}(af(x))=af'(x)$
ax^n	anx^{n-1}	$\dfrac{dy}{dx}=\dfrac{dy}{du}\times\dfrac{du}{dx}$
$\sin x$	$\cos x$	$\dfrac{dy}{dx}=1\div\dfrac{dx}{dy}$
$\cos x$	$-\sin x$	$\dfrac{d}{dx}(uv)=u\dfrac{dv}{dx}+v\dfrac{du}{dx}$
$\tan x$	$\sec^2 x$	$\dfrac{d}{dx}\left(\dfrac{u}{v}\right)=\dfrac{v\dfrac{du}{dx}-u\dfrac{dv}{dx}}{v^2}$

Any of these results can be used directly unless their derivation is asked for.

The next exercise gives practice in recognising the type of function and applying the most direct method to find its differential. Remember that some functions may fall into more than one category, so two rules may be needed. For example, $\dfrac{3x}{\sqrt{4x-1}}$ is a quotient and the denominator is a composite function. Remember also that you may be able to simplify some expressions before differentiating them.

Exercise 3.9b

1 Find $\dfrac{dy}{dx}$ when y is equal to:

(a) $\sin(x^2-1)$ (b) $\dfrac{\sqrt{1+x^2}}{x}$ (c) $(1+x^3)\sin x$ (d) $\dfrac{1+x}{1-x^2}$

2 Differentiate each of the following functions with respect to x.

(a) $\dfrac{3x}{\sqrt{4x-1}}$ (e) $(4-x^3)^5$ (i) $x^2\cos 2x$

(b) $\dfrac{\cos x}{\sin x}$ (f) $\dfrac{x-2}{x^2-4x+4}$ (j) $\left(\dfrac{x-1}{x+1}\right)^2$

(c) $\dfrac{1}{\cos x}$ (g) $\dfrac{x-2}{x^2+4x-4}$

(d) $(1+2x)\tan x$ (h) $\dfrac{\sin x}{x}$

3 Find $\dfrac{dy}{dx}$ in terms of the parameter when:

(a) $x=t^2+1,\ y=t+2$ (c) $x=\cos\theta,\ y=3\tan\theta$

(b) $x=\dfrac{1}{1+t},\ y=\dfrac{t-1}{t^2}$

Rate of increase

The gradient of a straight line, $y = mx + c$, is calculated from $\dfrac{\text{increase in } y}{\text{increase in } x}$ from one point on the line to another point on the line.
Therefore the gradient measures the rate at which y increases per unit increase in x, i.e. the rate of increase of y with respect to x.

$\dfrac{dy}{dx}$ gives the gradient of the tangent at a point on the curve $y = f(x)$,

so $\dfrac{dy}{dx}$ is a measure of the rate at which y is increasing with respect

to x at that point on the curve.

For example, at the point where

$x = 2$ on the curve $y = x^2$, $\dfrac{dy}{dx} = 2x = 4$

So where $x = 2$, y is increasing at the rate of 4 units for every unit increase in x.

Note that this is only true where $x = 2$ because the rate at which y changes varies as x varies.

At the point where $x = -2$, $\dfrac{dy}{dx} = -4$

The negative sign shows that y is *decreasing* at 4 units per unit increase in x.

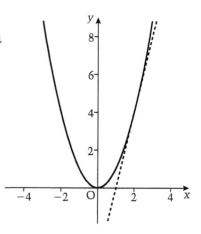

Connected rates of change

The chain rule is useful when we know the relationship between the variables y and x, we know the rate of change of y with respect to a variable u and we want to find the rate of change of x with respect to u.

Example

The equation of a curve is $y = 4 - \dfrac{1}{x}$

A point P is moving along the curve so that the y-coordinate is increasing at the constant rate of 0.01 units per second.
Find the rate at which the x-coordinate is increasing when $x = 1$

The rate at which y is increasing is $\dfrac{dy}{dt}$, so $\dfrac{dy}{dt} = 0.01$ where t seconds is the time.

$y = 4 - \dfrac{1}{x}$, therefore $\dfrac{dy}{dx} = \dfrac{1}{x^2}$

The rate of change of x with respect to t is $\dfrac{dx}{dt}$.

Using $\dfrac{dy}{dx} = \dfrac{dy}{dt} \times \dfrac{dt}{dx}$ gives $\dfrac{dt}{dx} = \dfrac{1}{x^2} \div 0.01 = \dfrac{100}{x^2}$

Now using $\dfrac{dx}{dt} = 1/\dfrac{dt}{dx}$ gives $\dfrac{dx}{dt} = \dfrac{x^2}{100}$

When $x = 1$, $\dfrac{dx}{dt} = 0.01$

Therefore when $x = 1$, x is increasing at the rate of 0.01 units per second.

Example

Air is leaking out of a spherical balloon at the constant rate of $0.3 \, cm^3$ per second.

Find the rate of change of the radius when the volume of the balloon is $36 \, cm^3$.

(The volume of a sphere is $V = \frac{4}{3}\pi r^3$) Give your answer correct to one significant figure.

When $V = 36$, $r = \dfrac{3}{\sqrt[3]{\pi}}$

We require $\dfrac{dr}{dt}$ when the volume is decreasing at the rate of $0.3 \, cm^3$ per second, i.e. when $\dfrac{dV}{dt} = -0.3$

From $V = \frac{4}{3}\pi r^3$, $\dfrac{dV}{dr} = 4\pi r^2$

Using $\dfrac{dV}{dr} = \dfrac{dV}{dt} \times \dfrac{dt}{dr}$ gives $4\pi r^2 = -0.3 \times \dfrac{dt}{dr} \Rightarrow \dfrac{dt}{dr} = -\dfrac{40}{3}\pi r^2$

Then $\dfrac{dr}{dt} = 1/\dfrac{dt}{dr} = -\dfrac{3}{40\pi r^2}$

When $r = \dfrac{3}{\sqrt[3]{\pi}}$, $\dfrac{dr}{dt} = -\dfrac{1}{120\pi^{\frac{1}{3}}} = -0.006$

The radius is decreasing at the rate of 0.006 cm per second.

Exercise 3.10

1   The area of a circular oil slick on a lake is increasing at the rate of $2 \, m^2$ per second.

Find the rate of change of the radius of the slick when the radius is 3 m. Give your answer correct to three significant figures.

2   The equation of a curve is $y = 2\sin\theta$. A point is moving along the curve so that θ is increasing at the constant rate of $\frac{1}{4}\pi$ radians per second. Find the rate of change of y when $\theta = \frac{7}{6}\pi$

3   A right circular cone has its axis vertical and its vertex downwards. It contains grain, which is pouring out of a hole in the vertex at the rate of $50 \, cm^3$ per second. The semi-vertical angle of the cone is $\frac{1}{6}\pi$. Find the rate of change of the height of grain in the cone when the radius of the circular surface of the grain is 2 m.

Increasing and decreasing functions

The value of $\dfrac{dy}{dx}$ at any point on a curve whose equation is $y = f(x)$ measures the rate at which y is increasing as x increases, i.e. $f'(x)$ measures the rate at which the function $f(x)$ is increasing with respect to x.

Consider, for example, the function given by $f(x) = x^3 - 3x + 2$

The graph shows the relationship between the curves $y = f(x)$ and $y = f'(x)$

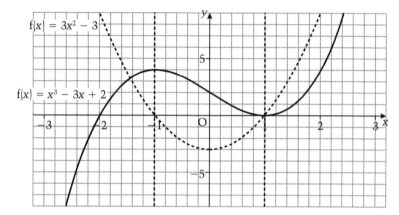

This graph shows that $f(x)$ is decreasing as x increases where $f'(x)$ is below the x-axis (i.e. $f'(x) < 0$), and $f(x)$ is increasing as x increases where $f'(x)$ is above the x-axis (i.e. $f'(x) > 0$).

**Therefore $f'(x) > 0$ when $f(x)$ is increasing
and $f'(x) < 0$ when $f(x)$ is decreasing.**

So, to determine whether a function is increasing or decreasing, we need to determine whether $f'(x)$ is positive or negative.

Example

Find the range of values of x for which the function
$$f(x) = 3x^2 - 2x + 4$$
is decreasing.

$f(x) = 3x^2 - 2x + 4$

$\Rightarrow \quad f'(x) = 6x - 2$

$\quad f'(x) < 0$ when $6x < 2$, i.e. when $x < \frac{1}{3}$

Therefore $f(x)$ is decreasing for values of x less than $\frac{1}{3}$

Example

Determine the range of values of x for which $f(x) = \dfrac{2x}{x^2 + 5x + 4}$ is increasing.

$$f(x) = \frac{2x}{x^2 + 5x + 4}$$

$$\Rightarrow \quad f'(x) = \frac{2(x^2 + 5x + 4) - 2x(2x + 5)}{(x^2 + 5x + 4)^2}$$

$$= \frac{-2x^2 + 8}{(x^2 + 5x + 4)^2}$$

$$= \frac{2(2 - x)(2 + x)}{[(x + 1)(x + 4)]^2}$$

$f'(x)$ is zero when $x = -2$ and $x = 2$ and $f'(x)$ is undefined when $x = -4$ and $x = -1$ so we need to investigate the sign of $f'(x)$ when $x < -4$, $-4 < x < -2$, $-2 < x < -1$, $-1 < x < 2$ and $x > 2$

Note that $[(x + 1)(x + 4)]^2 \geqslant 0$ for all values of x.

The table shows the sign of $f'(x)$ for the different ranges of x.

	$x < -4$	$-4 < x < -2$	$-2 < x < -1$	$-1 < x < 2$	$x > 2$
$2 - x$	+	+	+	+	−
$2 + x$	−	−	+	+	+
$[(x + 1)(x + 4)]^2$	+	+	+	+	+
$\dfrac{2(2 - x)(2 + x)}{[(x + 1)(x + 4)]^2}$	−	−	+	+	−

Therefore $f'(x) > 0$ for $-2 < x < -1$ and for $-1 < x < 2$, hence $f(x)$ is increasing for $-2 < x < -1$ and for $-1 < x < 2$

Exercise 3.11

1 Find the range of values of x for which the function given by
$$f(x) = x^3 - 4x^2 + 4x$$
is increasing.

2 Find the range of values of x for which the function given by
$$f(x) = \frac{2x}{(1 + x)^2}$$
is decreasing.

3.12   Stationary values

Learning outcomes

- To define and find stationary values

You need to know

- How to differentiate simple functions
- How to differentiate products, quotients and composite functions

Stationary values

We have seen in Topic 3.11 that when $f'(x)$ is positive, $f(x)$ is increasing as x increases, and that when $f'(x)$ is negative, then $f(x)$ is decreasing as x increases.

There may be points where $f(x)$ is neither increasing nor decreasing so $f'(x)$ is neither positive nor negative but is equal to zero.

The value of $f(x)$ at such a point is called a ***stationary value***.

Therefore $f'(x) = 0 \Rightarrow f(x)$ has a stationary value.

Consider the graph of $y = f(x)$.

At the points A, B and C, $f(x)$, and therefore y, is neither increasing nor decreasing. Therefore the values of y at these points are stationary values,

i.e. $\dfrac{dy}{dx} = 0 \Rightarrow y$ has a stationary value.

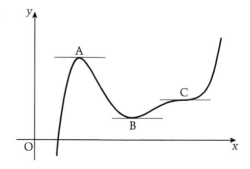

The points on the graph where y has a stationary value are called ***stationary points***.

At stationary points, the gradients of the tangents to the curve are zero, i.e. the tangents are parallel to the x-axis.

Therefore at a stationary point $\begin{cases} \textbf{y or f(x) has a stationary} \\ \textbf{value} \\ \dfrac{dy}{dx}, \textbf{or f}'(x), = 0 \\ \textbf{the tangent to the curve is} \\ \textbf{parallel to the axis} \end{cases}$

Example

Find the stationary values of the function given by
$f(x) = x^3 - 2x^2 + x - 1$

$f(x) = x^3 - 2x^2 + x - 1 \Rightarrow f'(x) = 3x^2 - 4x + 1$

At stationary values, $f'(x) = 0$,

$\Rightarrow \quad 3x^2 - 4x + 1 = 0$

$\Rightarrow \quad (3x - 1)(x - 1) = 0$

$\Rightarrow \quad x = \tfrac{1}{3} \text{ or } x = 1$

When $x = \tfrac{1}{3}$, $f(x) = \tfrac{1}{27} - \tfrac{2}{9} + \tfrac{1}{3} - 1 = -\tfrac{23}{27}$ and when $x = 1$, $f(x) = -1$

Therefore the stationary values of $f(x)$ are -1 and $-\tfrac{23}{27}$

Example

Show that the curve whose equation is $y = \dfrac{1}{x^2 - 2x + 1}$ has no stationary points.

$$y = \frac{1}{x^2 - 2x + 1} = \frac{1}{(x-1)^2} \quad \Rightarrow \quad \frac{dy}{dx} = \frac{-2}{(x-1)^3}$$

At stationary values, $\dfrac{dy}{dx} = 0$, but there are no values of x for

which $\dfrac{-2}{(x-1)^3} = 0$,

therefore the curve $y = \dfrac{1}{x^2 - 2x + 1}$ has no stationary values.

Example

Show that the curve whose equation is $y = \dfrac{x}{(x-1)^2}$ has only one stationary point and find it.

$$y = \frac{x}{(x-1)^2} \quad \Rightarrow \quad \frac{dy}{dx} = -\frac{x+1}{(x-1)^3}$$

The curve has a stationary point where $\dfrac{dy}{dx} = 0$,

i.e. where $-\dfrac{x+1}{(x-1)^3} = 0$

There is only one value of x for which $-\dfrac{x+1}{(x-1)^3} = 0$, i.e. $x = -1$

Therefore the curve has only one stationary point.

When $x = -1$, $y = -\frac{1}{4}$, so the stationary point is the point $\left(-1, -\frac{1}{4}\right)$.

Example

The curve $y = ax^2 + bx + 8$ has a stationary value of 5 when $x = 1$.

Find the values of a and b.

$$y = ax^2 + bx + 8 \quad \Rightarrow \quad \frac{dy}{dx} = 2ax + b$$

y has a stationary value of 5 when $x = 1$,

$\therefore \qquad 5 = a + b + 8 \quad \Rightarrow a + b = -3 \qquad [1]$

and $\quad 2a + b = 0 \qquad\qquad\qquad\qquad [2]$

Solving [1] and [2] simultaneously gives $a = 3$ and $b = -6$

Exercise 3.12

1 Find the stationary value of the function given by $f(x) = x^2 - 5x + 1$

2 Find the stationary value of the function given
   by $f(x) = x^3 - 6x^2 + 12x + 2$

3 Find the stationary points on the curve $y = \dfrac{x^2}{x+1}$

4 Show that there are no stationary points on the curve
   $$y = \frac{2x}{x^2 + 5x - 4}$$

- To define maximum and minimum points and points of inflexion
- To distinguish between stationary points
- To define the second differential of a function

- How to differentiate simple functions
- How to find stationary values
- How to determine whether a function is increasing or decreasing

Turning points

A curve $y = f(x)$ can have several stationary points and the shape of the curve close to one of these points belongs to three different types.

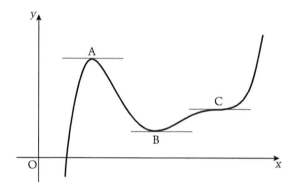

Moving along the curve in the positive direction of the x-axis:

1   Near the point A, the gradient of the curve goes from positive through zero to negative.

A is called a maximum turning point.

The value of y at A is called the maximum value of y (or of f(x)).

2   Near the point B, the gradient of the curve goes from negative through zero to positive.

B is called a minimum turning point.

The value of y at A is called the minimum value of y (or of f(x)).

3   At C the gradient is zero but the gradient does not change sign as a point on the curve moves through C. The curve does not turn here, but the sense of curvature does change from clockwise to anticlockwise.

A point on a curve at which the curvature changes from clockwise to anticlockwise (and vice versa) is called a point of inflexion.

There are two other points in the diagram where the sense of curvature changes, one between A and B and one between B and C. Therefore the gradient at a point of inflexion is not necessarily zero. However the gradient at a turning point *is* zero.

Note that the terms maximum and minimum do not mean greatest value and least value. They refer only to the behaviour of a function close to its stationary values.

Distinguishing between stationary values

There are three ways of distinguishing between stationary values.

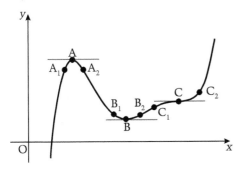

First method

Look at the points on either side of, and close to, the stationary values.

For A (a maximum value): y at $A_1 < y$ at A

                                      y at $A_2 < y$ at A

For B (a minimum value): y at $B_1 > y$ at B

                                      y at $B_2 > y$ at B

For C (a point of inflexion): y at $C_1 < y$ at C

                                      y at $C_2 > y$ at C

We can summarise these observations in a table:

	Maximum	Minimum	Inflexion
Values of y either side of the stationary value	Both smaller	Both larger	One smaller and one larger

Second method

For this method we look at the gradient on either side of, and close to, the stationary values.

For A (a maximum value): $\dfrac{dy}{dx}$ at $A_1 > 0$, $\dfrac{dy}{dx}$ at $A = 0$, $\dfrac{dy}{dx}$ at $A_2 < 0$

For B (a minimum value): $\dfrac{dy}{dx}$ at $B_1 < 0$, $\dfrac{dy}{dx}$ at $B = 0$, $\dfrac{dy}{dx}$ at $B_2 > 0$

For C (a point of inflexion): $\dfrac{dy}{dx}$ at $C_1 > 0$, $\dfrac{dy}{dx}$ at $C = 0$, $\dfrac{dy}{dx}$ at $C_2 > 0$

These results are summarised in the table:

	Maximum	Minimum	Inflexion	
Sign of $\dfrac{dy}{dx}$ either side, and at, a stationary value	+ 0 − / ─ \\	− 0 + \\ ─ /	+ 0 + / ─ /	or − 0 − \\ ─ \\

Third method

For A (a maximum value):

As a point moves along the curve through A, then as x increases $\frac{dy}{dx}$ goes from positive to negative

so $\frac{dy}{dx}$ *decreases* as x increases.

Therefore $\frac{dy}{dx}$ is a decreasing function at A.

For B (a minimum value):

As a point moves along the curve through B, then as x increases $\frac{dy}{dx}$ goes from negative to positive

so $\frac{dy}{dx}$ *increases* as x increases.

Therefore $\frac{dy}{dx}$ is an increasing function at A.

For C (a point of inflexion):

As a point moves along the curve through C, then as x increases $\frac{dy}{dx}$ goes from positive to zero then increases again to positive values.

Therefore $\frac{dy}{dx}$ itself has a stationary value at C.

Second derivative

The rate of change of $\frac{dy}{dx}$ is the derivative of $\frac{dy}{dx}$ with respect to x. This is called the *second derivative* of y and is denoted by $\frac{d^2y}{dx^2}$ or when $y = f(x)$, by $f''(x)$.

For example, when $y = x^3 - 2x^2$,

$\frac{dy}{dx} = 3x^2 - 4x$ and

$\frac{d^2y}{dx^2} = \frac{d}{dx}(3x^2 - 4x)$

$= 6x - 4$

Returning to the **third method** for distinguishing between stationary points, we can now express the observations above in terms of the second derivative.

At A, $\frac{dy}{dx}$ is a decreasing function therefore $\frac{d^2y}{dx^2} < 0$

At B, $\frac{dy}{dx}$ is an increasing function therefore $\frac{d^2y}{dx^2} > 0$

At C, $\frac{dy}{dx}$ has a stationary value, therefore $\frac{d^2y}{dx^2} = 0$

This is the easiest method to use. However $\frac{d^2y}{dx^2}$ can also be zero at a maximum or minimum value, so when $\frac{d^2y}{dx^2} = 0$ either the first or second method has to be used to distinguish between stationary values of y.

These results are summarised in the table:

	Maximum	**Minimum**
Sign of $\frac{d^2y}{dx^2}$ at a stationary value	Negative (or zero)	Positive (or zero)

Example

Find the stationary points on the curve $y = 3x^4 - 4x^3 + 2$ and distinguish between them.

$y = 3x^4 - 4x^3 + 2 \Rightarrow \frac{dy}{dx} = 12x^3 - 12x^2$

At stationary points, $\frac{dy}{dx} = 0 \quad \Rightarrow 12x^3 - 12x^2 = 0$

$\Rightarrow x^2(x - 1) = 0$

$\Rightarrow x = 0 \text{ or } x = 1$

When $x = 0$, $y = 2$ and when $x = 1$, $y = 1$

Therefore $(0, 2)$ and $(1, 1)$ are stationary points.

$\frac{d^2y}{dx^2} = 36x^2 - 24x$

When $x = 1$, $\frac{d^2y}{dx^2} = 36 - 24 > 0$

$\therefore (1, 1)$ is a minimum point.

When $x = 0$, $\frac{d^2y}{dx^2} = 0$ which is inconclusive, so we look at the signs of $\frac{dy}{dx}$ each side of $x = 0$

When $x = -\frac{1}{2}$, $\frac{dy}{dx} = 12(-\frac{1}{2})^3 - 12(-\frac{1}{2})^2 < 0$

When $x = \frac{1}{2}$, $\frac{dy}{dx} = \frac{12}{8} - \frac{12}{4} < 0$

Therefore $(0, 1)$ is a point of inflexion.

Exercise 3.13

Find the stationary points on the curve $y = f(x)$ and distinguish between them when $f(x)$ is:

(a) $x^3 - 3x^2 + 3x - 1$ **(b)** $x^3 - 2x^2 + x + 2$ **(c)** $(x - 2)^4$

Learning outcomes

- To use a variety of techniques to sketch curves

You need to know

- How to find stationary points and distinguish between them
- The meaning of an asymptote
- How to find limits
- How to solve rational inequalities
- How to convert an improper rational function to a proper rational function
- How to find the range of values that a rational function can take
- How to determine whether a function is increasing or decreasing

Features to look for when sketching curves

To sketch a curve whose shape is unknown, we can often find several features by observation and by calculation from the equation of the curve.

The main features to look for are:

- where the curve crosses the axes
- stationary points
- vertical and horizontal asymptotes
- the range of values of y
- the behaviour of y as $x \to \pm \infty$
- where the function is increasing or decreasing.

Not all of these features need to be considered for a particular curve. A picture of the curve can be built up by marking these features on a diagram as you find them.

Example

Sketch the curve whose equation is $y = \dfrac{x - 3}{x - 2}$

First find where the curve crosses the axes:

when $x = 0$, $y = 1\frac{1}{2}$ and when $y = 0$, $x = 3$

Therefore the curve goes through the points $(0, 1\frac{1}{2})$ and $(3, 0)$.

Now look for asymptotes:

as $x \to \pm \infty$, $1 - \dfrac{1}{x - 2} \to 1$ so the line $y = 1$ is an asymptote.

The value of y is undefined when $x = 2$ so the line $x = 2$ is an asymptote

and $\lim\limits_{x \to 2^+} \left(1 - \dfrac{1}{x - 2}\right) = -\infty$ and $\lim\limits_{x \to 2^-} \left(1 - \dfrac{1}{x - 2}\right) = \infty$

These findings are shown on the diagram.

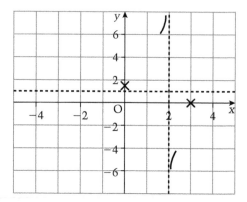

Next investigate stationary points:

$$y = \frac{x - 3}{x - 2}$$

$$= 1 - \frac{1}{x - 2}$$

$$= 1 - (x - 2)^{-1}$$

$$\Rightarrow \frac{dy}{dx} = \frac{1}{(x - 2)^2}$$

There are no values of x for which $\frac{dy}{dx} = 0$ so there are no stationary values.

Also $\frac{dy}{dx} > 0$ for all values of x except $x = 2$, therefore y is increasing as x increases.

Check the range of values of y:

$$y = \frac{x - 3}{x - 2} \Rightarrow x = \frac{2y - 3}{y - 1}$$ so x has real values except when $y = 1$, therefore the curve does not cross the line $y = 1$

We now have enough information to sketch the curve.

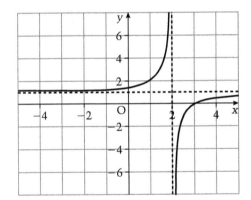

Example

Sketch the curve $y = \frac{(x - 1)^2}{x(2x - 1)}$

Check for intercepts on the axes:

when $x = 0$, y is undefined. Therefore the curve does not cross the y-axis.

when $y = 0$, $x = 1$. This is a repeated root, therefore the curve touches the x-axis at $(1, 0)$.

(This must be a stationary point.)

Check for asymptotes:

y is undefined when $x = 0$ and $x = \frac{1}{2}$,

therefore the lines $x = 0$ and $x = \frac{1}{2}$ are vertical asymptotes.

$$y = \frac{(x - 1)^2}{2x^2 - x}$$

$$= \frac{x^2 - 2x + 1}{2x^2 - x}$$

$$= \frac{1}{2} - \frac{3x - 2}{4x^2 - 2x}$$

$$= \frac{1}{2} - \frac{3x - 2}{2x(2x - 1)}$$

As $x \to +\infty$, $\left(\frac{1}{2} - \frac{3x - 2}{2x(2x - 1)}\right) \to \frac{1}{2}$ from below

and as $x \to -\infty$, $\left(\frac{1}{2} - \frac{3x - 2}{2x(2x - 1)}\right) \to \frac{1}{2}$ from above.

Therefore $y = \frac{1}{2}$ is a horizontal asymptote.

When $y = \frac{1}{2}$,

$$\frac{1}{2} = \frac{1}{2} - \frac{3x - 2}{2x(2x - 1)} \Rightarrow \frac{3x - 2}{2x(2x - 1)} = 0 \Rightarrow x = \frac{2}{3}$$

Therefore the curve crosses the line $y = \frac{1}{2}$ where $x = \frac{2}{3}$

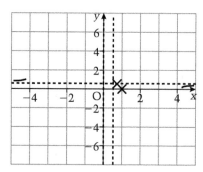

This diagram shows the features found so far.

As the curve crosses the asymptote $y = \frac{1}{2}$ only once, and does not cross the vertical asymptotes, it is clear that the point $(1, 0)$ is a minimum point.

We can also deduce that $y > \frac{1}{2}$ when $x < 0$

and $y > 0$ when $x > \frac{1}{2}$

We now check for other stationary points.

$$y = \frac{(x - 1)^2}{x(2x - 1)} \Rightarrow \frac{dy}{dx} = \frac{2(x - 1)(2x^2 - x) - (x - 1)^2(4x - 1)}{x^2(2x - 1)^2}$$

$$= \frac{(x - 1)(3x - 1)}{x^2(2x - 1)^2}$$

$\frac{dy}{dx} = 0$ when $x = 1$ and $\frac{1}{3}$

Therefore y has stationary points where $x = 1$ and where $x = \frac{1}{3}$

When $x = 1$, $y = 0$ and when $x = \frac{1}{3}$, $y = \dfrac{\left(-\frac{2}{3}\right)^2}{\left(\frac{1}{3}\right)\left(-\frac{1}{3}\right)} = -4$

We will use the values of x either side of the stationary points to determine their nature.

(The values chosen must be close to the value of x at the stationary point and the curve must be continuous between those values.)

x	$\frac{1}{4}$	$\frac{1}{3}$	$\frac{5}{12}$	$\frac{3}{4}$	1	2
y	< -4	-4	< -4	> 0	0	> 0

Therefore $\left(\frac{1}{3}, -4\right)$ is a maximum point and $(1, 0)$ is a minimum point.

We could also check the range of values that y can take and the values of x for which y is increasing and decreasing, but we now have enough information to sketch the curve.

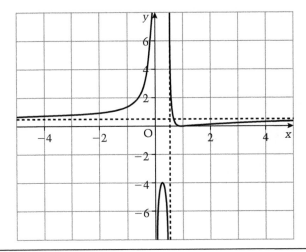

Exercise 3.14

Sketch the graphs of the following curves.

1 $y = \dfrac{1}{2 - x}$

2 $y = \dfrac{x}{2 - x}$

3 $y = \dfrac{x^2}{(2 - x)^2}$

Equations of tangents and normals

We know that $\dfrac{dy}{dx}$ represents the gradient function of a curve whose equation is $y = f(x)$

We can therefore find the gradient of the curve, and hence the gradient of the tangent to the curve, at any given point on the curve.

When $x = a$, $y = f(a)$ and the gradient at that point is $f'(a)$.

Therefore the equation of the tangent to the curve $y = f(x)$ at the point $(a, f(a))$ is given by

$$y - f(a) = f'(a)(x - a)$$

The **normal** to a curve is a line perpendicular to the tangent and through the point of contact of the tangent. Therefore at the point $(a, f(a))$ the gradient of the normal is $-\dfrac{1}{f'(a)}$, and the equation of the normal to the curve at the point $y = f(x)$ at the point $(a, f(a))$ is given by

$$y - f(a) = -\frac{1}{f'(a)}(x - a)$$

For example, when $y = \sqrt{x + 1}$, $\dfrac{dy}{dx} = \dfrac{1}{2\sqrt{x + 1}}$

When $x = 3$, $\dfrac{dy}{dx} = \dfrac{1}{4}$, so the gradient of the curve at the point where $x = 3$ is $\dfrac{1}{4}$

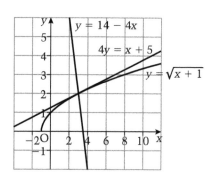

When $x = 3$, $y = 2$, therefore the line through $(3, 2)$ with gradient $\frac{1}{4}$ is a tangent to the curve at the point $(3, 2)$.

The equation of this tangent is $y - 2 = \frac{1}{4}(x - 3)$

$$\Rightarrow \quad 4y = x + 5$$

The equation of the normal at this point

is $\quad y - 2 = -4(x - 3)$

$$\Rightarrow \quad y = 14 - 4x$$

Equations of tangents and normals for curves whose equations are parametric

When the equation of a curve is given as $x = f(t)$ and $y = g(t)$, a point on the curve can be expressed as $(f(t), g(t))$ and the gradient function is given by $\dfrac{dy}{dx} = \dfrac{dy}{dt} \div \dfrac{dx}{dt} = \dfrac{g'(t)}{f'(t)}$

Therefore the equation of a tangent to the curve can be given as

$$y - g(t) = \frac{g'(t)}{f'(t)}(x - f(t))$$

Then any particular value of t gives the equation of the tangent at that point.

Similarly, the equation of a normal to the curve can be given as

$$y - g(t) = -\frac{f'(t)}{g'(t)}(x - f(t))$$

Example

Find the equation of the tangent and the normal to the curve

$x = 3\cos\theta$, $y = 4\sin\theta$, at the point where $\theta = \frac{\pi}{6}$

$$f(\theta) = 3\cos\theta \text{ and } g(\theta) = 4\sin\theta$$
$$\Rightarrow \quad f'(\theta) = -3\sin\theta \text{ and } g'(\theta) = 4\cos\theta$$

so the equation of a tangent to this curve is

$$y - 4\sin\theta = -\frac{4\cos\theta}{3\sin\theta}(x - 3\cos\theta)$$

(This is a general equation; it gives the equation of a tangent to the curve at any point on the curve.)

Therefore the equation of the tangent at the point where $\theta = \frac{\pi}{6}$ is given by

$$y - 4\left(\tfrac{1}{2}\right) = -\frac{4\left(\frac{\sqrt{3}}{2}\right)}{3\left(\frac{1}{2}\right)}\left(x - 3\left(\tfrac{\sqrt{3}}{2}\right)\right) \quad \Rightarrow \quad y - 2 = -\frac{4\sqrt{3}}{3}\left(x - \tfrac{3\sqrt{3}}{2}\right)$$

The equation of a normal to this curve is given by

$$y - 4\sin\theta = \frac{3\sin\theta}{4\cos\theta}(x - 3\cos\theta)$$

(Again this is a general equation giving the equation of a normal at any point on the curve.)

So the equation of the normal at the point where $\theta = \frac{\pi}{6}$ is given by

$$y - 4\left(\tfrac{1}{2}\right) = \frac{3\left(\frac{1}{2}\right)}{4\left(\frac{\sqrt{3}}{2}\right)}\left(x - 3\left(\tfrac{\sqrt{3}}{2}\right)\right) \quad \Rightarrow \quad y - 2 = \frac{3}{4\sqrt{3}}\left(x - \tfrac{3\sqrt{3}}{2}\right)$$

Exercise 3.15

1 Find the equation of the tangent and normal to the curve whose equation is $y = \frac{2}{(x-3)}$ at the point where $x = 7$

2 Find the equations of the tangent and normal to the curve whose equations are $x = \frac{1}{1+t}$, $y = t - 1$

Reversing differentiation

When x^2 is differentiated with respect to x the derivative is $2x$.

Therefore when the derivative of an unknown function is $2x$ then the unknown function could be x^2.

This process of finding a function from its derivative, which reverses the operation of differentiating, is called **integration**.

The constant of integration

We know that $2x$ is the derivative of x^2, but it is also the derivative of $x^2 + 5,\ x^2 - 3$

In fact, $2x$ is the derivative of $x^2 + c$, where c is any constant.

Therefore the result of integrating $2x$ is not a unique function but is of the form $x^2 + c$ where c is any constant.

c is called the **constant of integration**.

$x^2 + c$ is called the **integral** of $2x$ with respect to x and is written as $\int 2x\,dx = x^2 + c$

where $\int \ldots dx$ means 'the integral of ... with respect to x'.

Integrating *any* function reverses the process of differentiating, so for any function $f(x)$ we have

$$\int f'(x)\,dx = f(x) + c$$

For example, differentiating x^3 with respect to x gives $3x^2$ so $\int 3x^2\,dx = x^3 + c$ and it follows that $\int x^2\,dx = \frac{1}{3}x^3 + c$

(We do not need to write $\frac{1}{3}c$ in the second form, as c represents *any* constant in either expression.)

Now we know that the derivative of x^{n+1} is $(n+1)x^n$ so

$$\int x^n\,dx = \frac{1}{n+1}x^{n+1} + c$$

Therefore to integrate a power of x, *increase* that power by 1 and *divide* by the new power.

This rule can be used to integrate any power of x *except* -1.

For example, $\int 6x^3\,dx = \frac{6}{4}x^4 + c = \frac{3}{2}x^4 + c$,

$\int 3\,dx = \int 3x^0\,dx = 3x + c$ and $\int 4x^{-3}\,dx = \frac{4}{-2}x^{-2} + c = -2x^{-2} + c$

Exercise 3.16a

Find the following integrals.

(a) $\int 5x\,dx$ **(c)** $\int 6\,dx$ **(e)** $\int x^{-2}\,dx$

(b) $\int 4x^7\,dx$ **(d)** $\int 4x^3\,dx$ **(f)** $\int 5x^{-6}\,dx$

Families of curves

When $\dfrac{dy}{dx} = 2x,$ then $y = \int 2x\,dx = x^2 + c$

Therefore the equation $y = x^2 + c$ represents a family of curves.

Each value of c gives a different member of the family.

The graph shows some members of this family.

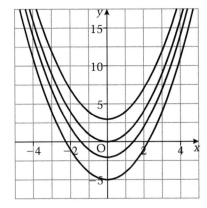

To find the equation of a particular member, we need to know a point on the curve.

For example, if $y = \int 3x^2\,dx$

then $\qquad\quad y = x^3 + c$

If we also know that $(2, 5)$ is a point on the curve we can find the value of c and hence the equation of the particular curve.

When $x = 2$ and $y = 3,$ $y = x^3 + c \;\Rightarrow\; 3 = 8 + c \;\Rightarrow\; c = -5$

Therefore the equation of the curve is $y = x^3 - 5$

Exercise 3.16b

1 Find the equation of the curve that goes through $(1, 5)$ and for which $y = \int 6x^2\,dx$

2 Find the equation of the curve that goes through $(-1, 2)$ and for which $y = \int 9x^{-4}\,dx$

3.17 Integration of sums and differences of functions

Learning outcomes

- To define the integral of sums and differences of functions

You need to know

- The differentials of simple functions
- The meaning of integration as the reverse of differentiation
- The meaning of the constant of integration
- How to solve a pair of simultaneous equations
- The effect on the equation of a curve by a translation parallel to the y-axis

Integration of a sum or difference of functions

When $y = f(x) + g(x)$ we know that $\dfrac{dy}{dx} = f'(x) + g'(x)$

Therefore it follows that $\int (f'(x) + g'(x))\, dx = f(x) + g(x) + c$

For example, $\int (\cos x + \sin x)\, dx = \sin x - \cos x + c$

When $y = f(x) - g(x)$ we know that $\dfrac{dy}{dx} = f'(x) - g'(x)$

Therefore it follows that $\int (f'(x) - g'(x))\, dx = f(x) - g(x) + c$

For example, $\int (2x - \sin x)\, dx = x^2 + \cos x + c$ and
$\int (1 - 2x)\, dx = x - x^2 + c$

Integration of a multiple of a function

When $y = af(x)$ we know that $\dfrac{dy}{dx} = af'(x)$

Therefore $\int af'(x) = af(x) + c$

For example, $\int 6x^2\, dx = \int 3(2x^2)\, dx = 2x^3 + c$

To summarise,

$$\int (f(x) \pm g(x))\, dx = \int f(x)\, dx \pm \int g(x)\, dx \quad \text{and} \quad \int (af(x))\, dx = a\int f(x)\, dx$$

Note that $\int (f(x) \times/\div g(x))\, dx$ is *not* equal to $\int f(x)\, dx \times/\div \int g(x)\, dx$

Example

Find $\int (x^3 - 4\cos x)\, dx$

$\int (x^3 - 4\cos x)\, dx = \int x^3\, dx - 4\int \cos x\, dx$
$\qquad\qquad\qquad\qquad = \tfrac{1}{4}x^4 - 4\sin x + c$

Exercise 3.17a

Find the following integrals:

(a) $\int 9x^2\, dx$

(b) $\int x(3x^3 - 4)\, dx$ (Hint: Multiply out the bracket.)

(c) $\int (5\sin x - 6\cos x)\, dx$

Example

The equation of a curve is such that $\dfrac{d^2y}{dx^2} = 6x - 14$. Two points on the curve have coordinates $(1, -2)$ and $(-1, -12)$. Find the equation of the curve.

$\dfrac{d^2y}{dx^2}$ is the second derivative of y with respect to x, so we need to integrate twice.

The first integral gives $\dfrac{dy}{dx}$

$$\dfrac{d^2y}{dx^2} = 6x - 14 \quad \Rightarrow \quad \dfrac{dy}{dx} = \int(6x - 14)\,dx \quad \Rightarrow \quad \dfrac{dy}{dx} = 3x^2 - 14x + c$$

We need to integrate again to find an expression for y so we need to introduce another unknown constant.

$$y = \int(3x^2 - 14x + c)\,dx \quad \Rightarrow \quad y = x^3 - 7x^2 + cx + d$$

We can now use the coordinates of the points on the curve to find the values of c and d.

When $\quad x = 1, y = -2, \quad \Rightarrow \quad -2 = -6 + c + d$, i.e. $c + d = 4 \qquad$ [1]

When $\quad x = -1, y = -12 \quad \Rightarrow \quad -12 = -8 - c + d$, i.e. $c - d = 4 \qquad$ [2]

Solving [1] and [2] simultaneously gives $c = 4$ and $d = 0$

Therefore $y = x^3 - 7x^2 + 4x$

Example

The equations of a family of curves is given by $\dfrac{dy}{dx} = 4x - 3$

Sketch the graphs of two members of this family.

$$\dfrac{dy}{dx} = 4x - 3 \quad \Rightarrow \quad y = \int(4x - 3)\,dx = 2x^2 - 3x + c$$

We can use any value of c to get one member of the family.
The simplest is $c = 0$

This gives $y = 2x^2 - 3x = x(2x - 3)$ which is a parabola that passes through O and $\left(\frac{3}{2}, 0\right)$.

Then any other value of c translates the curve by c units up the y-axis.

The sketch shows $y = 2x^2 - 3x$ and $y = 2x^2 - 3x + 2$

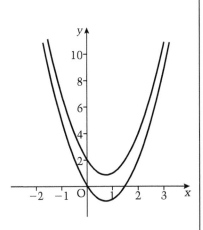

Exercise 3.17b

1. The equation of a curve is such that $\dfrac{dy}{dx} = 3\cos x$ and the curve passes through the point $\left(\frac{\pi}{2}, 4\right)$.
   Find the equation of the curve.

2. A curve passes through the points $(0, 1)$ and $(1, 6)$. The equation of the curve is such that $\dfrac{d^2y}{dx^2} = 6x$
   Find the equation of the curve.

Learning outcomes

■ To use substitution to integrate functions

You need to know

■ The chain rule
■ The differentials of simple functions
■ The integrals of simple functions

Integration using substitution

When we use the chain rule to differentiate a composite function the result is often a product of functions. For example, when $y = (x^3 - 4x)^4$, then with $u = x^3 - 4x$ so that $y = u^4$, the chain rule gives

$$\frac{dy}{dx} = 4u^3(3x^2 - 4) = 4(3x^2 - 4)(x^3 - 4x)^3$$

In general, if $u = f(x)$ and $y = g(u)$ then using the chain rule gives

$$\frac{dy}{dx} = \frac{dy}{du} \times \frac{du}{dx} = g'(u)\frac{du}{dx}$$

Therefore, using integration as the reverse of differentiation

$$\int g'(u) \frac{du}{dx} dx = g(u) + c \qquad [1]$$

Now $\quad \int g'(u) du = g(u) + c \qquad [2]$

Comparing [1] and [2] gives

$$\int g'(u) du = g(u) + c$$

Replacing $g'(u)$ by $f(u)$ gives

$$\int f(u) \frac{du}{dx} dx = \int f(u) du$$

Therefore $\qquad\qquad \ldots \frac{du}{dx} dx \equiv \ldots du$

This means that integrating $\left((a \text{ function of } u) \frac{du}{dx}\right)$ with respect to x is *equivalent* to integrating (the same function of u) with respect to u.

This means that the relationship $\ldots \dfrac{du}{dx} dx \equiv \ldots du$ is not an equation or an identity – it is a pair of equivalent operations.

For example, to find $\int 3x^2(x^3 + 4)^4 dx$ we can use the substitution $u = x^3 + 4$

This gives $\int 3x^2(x^3 + 4)^4 dx = \int 3x^2 u^4 dx$

Then as $\dfrac{du}{dx} = 3x^2$, $\ldots \dfrac{du}{dx} dx \equiv \ldots du$ becomes $\ldots 3x^2 dx = \ldots du$

$\therefore \quad \int 3x^2 u^4 dx = \int u^4 du \qquad\qquad$ Since $\ldots 3x^2 dx = \ldots du$

$$= \tfrac{1}{5} u^5 + c$$

$$= \tfrac{1}{5} (x^3 + 4)^5 + c$$

Example

Use the substitution $u = \sin x$ to find $\int \cos x \sin^2 x \, dx$

$u = \sin x \quad \Rightarrow \quad \dfrac{du}{dx} = \cos x,$

$\therefore \quad \ldots \dfrac{du}{dx} \, dx \equiv \ldots du \quad \Rightarrow \quad \ldots \cos x \, dx = \ldots du$

Therefore $\int \cos x \sin^2 x \, dx = \int u^2 \, du$

$$= \tfrac{1}{3} u^3 + c$$

$$= \tfrac{1}{3} \sin^3 x + c$$

Example

Use the substitution $u = \sqrt{1 + x}$ to find $\int (x\sqrt{1 + x}) \, dx$

$u = \sqrt{1 + x} \quad \Rightarrow \quad \dfrac{du}{dx} = \tfrac{1}{2}(1 + x)^{-\frac{1}{2}}$ and $u^2 - 1 = x$

$\therefore \quad \ldots \dfrac{du}{dx} \, dx \equiv \ldots du \quad \Rightarrow \quad \ldots \tfrac{1}{2}(1 + x)^{-\frac{1}{2}} \, dx = \ldots du$

$\quad \Rightarrow \quad \ldots dx = \ldots 2\sqrt{1 + x} \, du = \ldots 2u \, du$

Hence $\int (x\sqrt{1 + x}) \, dx = \int (u^2 - 1)(u)(2u) \, du = \int (2u^4 - 2u^2) \, du$

$$= \tfrac{2}{5} u^3 - \tfrac{2}{3} u^3 + c$$

$$= \tfrac{2}{5}(1 + x)^{\frac{5}{2}} - \tfrac{2}{3}(1 + x)^{\frac{3}{2}} + c$$

$$= \tfrac{2}{15}(1 + x)^{\frac{3}{2}}(3(1 + x) - 5) + c$$

$$= \tfrac{2}{15}(3x - 2)\sqrt{(1 + x)^3} + c$$

Exercise 3.18

1 Use the substitution $u = 2x$ to find $\int \cos 2x \, dx$

2 Use the substitution $u = x^2 + 1$ to find $\int 6x \, (x^2 + 1)^3 \, dx$

3 Use the substitution $u = x - 4$ to find $\int (x + 1)(x - 4)^6 \, du$

4 Use the substitution $u = \tan \theta$ to find $\int \sec^2 \theta \tan^2 \theta \, d\theta$

$\qquad\qquad\qquad\qquad$ (Hint: $\dfrac{d}{d\theta}(\tan \theta) = \sec^2 \theta$)

Calculus

The topics in Section 3 of this book all come under the umbrella name of *calculus*.

Calculus is the study of limits, derivatives and integrals; it basically studies how things change when the rate of change varies.

Our modern world would look very different without calculus. It is used to model situations that involve change and to predict what will happen when change takes place. It has applications in almost everything we do today, from the exploration of space to the development of the tiny microchips found in electronic devices. Apart from the obvious scientific uses, calculus covers a range of disciplines such as economics and graphic design.

The area under a curve

We will now look at how the fundamental theorem of calculus is derived.

Consider the area A in the diagram below enclosed by the curve $y = f(x)$, the x- and y-axes and the vertical line through a value of x.

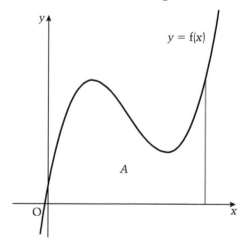

We can find an approximate value for this area by dividing it into vertical strips as shown in the next diagram.

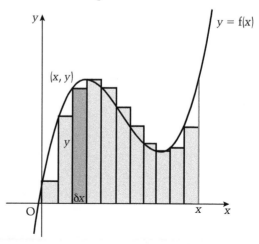

The area of a typical strip of height y and width δx is $y\,\delta x$.

The area is then approximately the sum of all the strips.

We write this as $A \approx \sum\limits_{x=0}^{x=a} y\,\delta x$ where $\sum\limits_{x=0}^{x=a} y\,\delta x$ means the sum of all

values of $y\,\delta x$ between $x = 0$ and $x = a$

This approximation improves as δx gets smaller, so we can write

$$A = \lim_{\delta x \to 0} \left(\sum_{x=0}^{x=a} y\,\delta x \right)$$

Considered this way, the area under a curve is the process that involves putting together (i.e. integrating) different elements.

We now look at a different approach to finding A.

If A is the area enclosed by the curve $y = f(x)$, the x- and y-axes and the vertical line through the point (x, y) on the curve, then a small increase in x, δx, gives a corresponding small increase in A, δA.

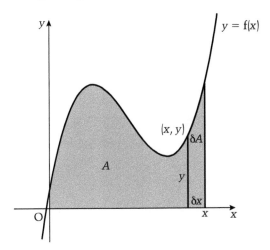

δA is approximately equal to the area of the rectangle of height y and width δx,

i.e. $\delta A \approx y\,\delta x$

$\therefore \quad \dfrac{\delta A}{\delta x} \approx y$ and this approximation gets better as $\delta x \to 0$

Now $\lim\limits_{\delta x \to 0} \dfrac{\delta A}{\delta x} = \dfrac{dA}{dx}$

hence $\dfrac{dA}{dx} = y$

This gives the connection between finding A as a summation process and the differential of A with respect to x, i.e. A is the reverse of a differential, hence

$$A = \int y\,dx$$

This is called the ***fundamental theorem of calculus***.

Definite integration

We know that the area between a curve $y = f(x)$, the x- and y-axes and the line through a value of x is given by $\int y\,dx$, i.e. by $\int f(x)\,dx$.

To find this area up to the line $x = b$, we find $\int f(x)\,dx$ and substitute b for x.

To find this area up to the line $x = a$, we find $\int f(x)\,dx$ and substitute a for x.

Then the area between the curve, the x-axis and the lines $x = a$ and $x = b$ is the difference between these two calculations.

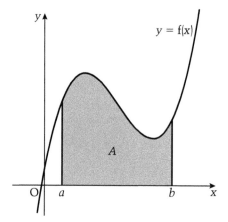

For example, when $y = x^2 + 2$, $\int y\,dx = \int(x^2 + 2)\,dx$
$$= \tfrac{1}{3}x^3 + 2x + c$$

Then the area up to $x = 1$ is $\tfrac{1}{3}(1)^3 + 2(1) + c = 2\tfrac{1}{3} + c$

and the area up to $x = 2$ is $\tfrac{1}{3}(2)^3 + 2(2) + c = 6\tfrac{2}{3} + c$

Therefore the area between the curve $y = x^2 + 2$, the x-axis and $x = 1$ and $x = 2$ is

$\left(6\tfrac{2}{3} + c\right) - \left(2\tfrac{1}{3} + c\right) = 4\tfrac{1}{3}$. Notice that the unknown constant disappears.

This process is called **definite integration**. It is written as $\int_a^b f(x)\,dx$ and it means the value of the integrand when $x = b$ minus the value of the integrand when $x = a$. The values a and b are called the **limits of the definite integral**.

We can write the example above more precisely as

$$\int_1^2 (x^2 + 2)\,dx = \left[\tfrac{1}{3}x^3 + 2x\right]_1^2 = \left(\tfrac{8}{3} + 4\right) - \left(\tfrac{1}{3} + 2\right) = 4\tfrac{1}{3}$$

The unknown constant of integration does not need to be included because it cancels out. The square brackets show the values of x to be substituted and the order in which they are to substituted (top one first). The second value is then subtracted from the first value.

$\int_a^b f(x)\,dx$ can be found only when $f(x)$ is continuous from $x = a$ to $x = b$

For example $\int_{-1}^1 \left(\dfrac{1}{x}\right)\,dx$ cannot be found because $\dfrac{1}{x}$ is undefined when $x = 0$

Example

Evaluate $\int_0^{\frac{\pi}{2}} \sin x \, dx$

$$\int_0^{\frac{\pi}{2}} \sin x \, dx = \left[-\cos x \right]_0^{\frac{\pi}{2}}$$

$$= -\cos \frac{\pi}{2} - (-\cos 0)$$

$$= 0 - (-1) = 1$$

Exercise 3.20a

Evaluate the following definite integrals:

(a) $\int_1^2 4x^3 \, dx$

(c) $\int_{-1}^1 (x - 1)^2 \, dx$

(b) $\int_2^4 (2x - 3) \, dx$

(d) $\int_1^3 (x - 5) \, dx$

Definite integration using substitution

In general when we use a substitution $u = g(x)$ to find an integral, we need to substitute back to give the answer in terms of x. However, we do not need to do this for a definite integral because we can use the values of x given as the limits to find the corresponding values of u.

Example

Use the substitution $u = x^3 + 2$ to evaluate $\int_0^2 x^2 (x^3 + 2)^4 \, dx$

$$u = x^3 + 2 \quad \Rightarrow \quad \ldots du = \ldots 3x^2 \, dx$$

and $\begin{cases} x = 0 & \Rightarrow \quad u = 2 \\ x = 2 & \Rightarrow \quad u = 10 \end{cases}$

$$\therefore \int_0^2 x^2 (x^3 + 2)^4 \, dx = \int_2^{10} \frac{1}{3} u^4 \, du$$

$$= \left[\frac{1}{15} u^5 \right]_2^{10}$$

$$= \frac{100\,000}{15} - \frac{32}{15} = \frac{99\,968}{15}$$

Exercise 2.20b

1 Use the substitution $u = x - 1$ to evaluate $\int_1^3 7(x - 1)^6 \, dx$

2 Use the substitution $u = \sin x$ to evaluate $\int_0^{\frac{\pi}{4}} \cos x (1 + \sin x)^2 \, dx$

3 Use the substitution $u^2 = 1 + x^2$ to evaluate $\int_0^1 x\sqrt{1 + x^2} \, dx$

Learning outcomes

- To calculate the area under a curve

You need to know

- How to integrate simple functions
- How to evaluate a definite integral
- How to sketch curves

Calculating the area under a curve

From Topics 3.19 and 3.20 we now know that the area between the curve $y = f(x)$, the x-axis and the lines $x = a$ and $x = b$ shown in the diagram is

the value of $\displaystyle\int_a^b f(x)\,dx$.

For example, the area between $y = x^2$, the x-axis and $x = 1$ and $x = 2$ is

given by $\displaystyle\int_1^2 x^2\,dx = \left[\tfrac{1}{3}x^3\right]_1^2 = \tfrac{8}{3} - \tfrac{1}{3} = \tfrac{7}{3}$

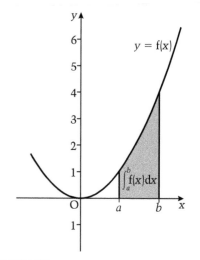

Example

Find the area enclosed by the curve $y = (1 - x)(2 + x)$ and the x-axis.

The curve is a parabola which cuts the x-axis at $x = -2$ and $x = 1$

The sketch shows the area required. (It is always sensible to draw a sketch.)

$$A = \int_{-2}^{1} (1 - x)(2 + x)\,dx$$

$$= \int_{-2}^{1} (2 - x - x^2)\,dx = \left[2x - \tfrac{1}{2}x^2 - \tfrac{1}{3}x^3\right]_{-2}^{1}$$

$$= \left(2 - \tfrac{1}{2} - \tfrac{1}{3}\right) - \left(-4 - 2 + \tfrac{8}{3}\right)$$

$$= 4\tfrac{1}{2}$$

The area is $4\tfrac{1}{2}$ square units.

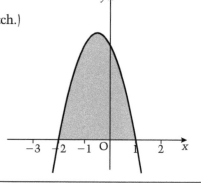

Exercise 3.21a

Find the area enclosed by each of the following curves, the x-axis and the lines given.

1   $y = 4x^3$, $x = 0$, $x = 2$

2   $y = x^2 + 1$, $x = -1$, $x = 1$

3   $y = \sqrt{x}$, $x = 1$, $x = 4$

Find the area enclosed by each of the following curves and the x-axis.

4   $y = 1 - x^2$

5   $y = x(1 - x)$

The area between a curve and the y-axis

The diagram shows the area between the curve $y = x^2 + 1$, the y-axis and the line $y = 10$.

We can find this area in two different ways.

First method

This uses the fact that the area required is equal to the area of the bounding rectangle minus the area between the curve and the x-axis.

When $y = 10$, $x = 3$ so the area shown is equal to

(area of the rectangle bounded by $x= 0$, $y = 0$, $x = 3$ and $y = 10$)

    $-$ (area between the curve, the x-axis, $x = 0$ and $x = 3$)

Therefore the area required $= 30 - \int_0^3 (x^2 + 1)\, dx$

$$= 30 - \left[\tfrac{1}{3}x^3 + x\right]_0^3$$

$$= 30 - \left\{\left(\tfrac{27}{3} + 3\right) - (0)\right\} = 30 - 12 = 18$$

The area is 18 square units.

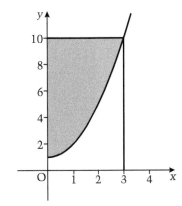

Second method

This uses a direct approach.

Consider a horizontal strip of length x and width δy.

The area of this strip is approximately $x\,\delta y$

The area required $= \lim\limits_{\delta y \to 0} \left(\sum\limits_{y=1}^{y=10} x\,\delta y\right)$

We know that this limit is equal to $\int_1^{10} x\, dy$

Now $\int_1^{10} x\, dy$ means integrate x with respect to y so we need to find x in terms of y.

From $y = x^2 + 1$, $x = \sqrt{y - 1}$

so the required area is $\int_1^{10} \left(\sqrt{y - 1}\right) dy = \int_0^9 u^{\frac{1}{2}}\, du = \left[\tfrac{2}{3} u^{\frac{3}{2}}\right]_0^9 = 18$ square units   Using the substitution $u = y - 1$

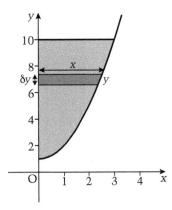

Exercise 3.21b

1 Find the area enclosed by the curve $y = \sqrt{x}$, the y-axis and the line $y = 3$

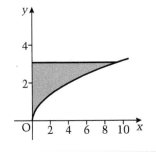

2 Find the area enclosed by the curve $y = x^2$, the y-axis and the line $y= 4$

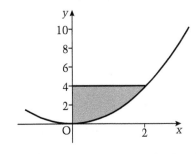

3.22 Area below the x-axis and area between two curves

Learning outcomes

- To calculate areas of curves below the x-axis
- To calculate areas between curves

You need to know

- How to integrate simple functions
- How to find the area below a curve and above the x-axis
- How to sketch curves
- How to find the points of intersection of two curves

The area between a curve and the x-axis that is below the axis

The definite integral

$$\int_1^2 x^3 \, dx = \left[\tfrac{1}{4}x^4\right]_1^2$$

$$= 4 - \tfrac{1}{4} = 3\tfrac{3}{4}$$

This can be interpreted as the area between the curve $y = x^3$, the x-axis and the lines $x = 1$ and $x = 2$

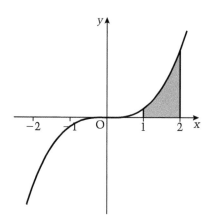

The definite integral

$$\int_{-2}^{-1} x^3 \, dx = \left[\tfrac{1}{4}x^4\right]_{-2}^{-1}$$

$$= \tfrac{1}{4} - 4 = -3\tfrac{3}{4}$$

If we look at the diagram, the area between the curve $y = x^3$, the x-axis and the lines $x = -2$ and $x = -1$ is equal to the area found above, i.e. $3\tfrac{3}{4}$ square units.

(The curve has rotational symmetry about O.)

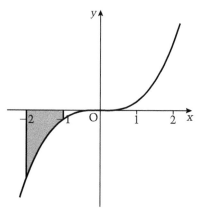

But the integral $\int_{-2}^{-1} x^3 \, dx$ is negative. This is because the value of y that gives the length of a vertical strip is negative, so $\int_a^b y \, dx$ will be negative when y is negative for $a \leqslant x \leqslant b$.

This means you need to be careful when finding the area between a curve and the x-axis when part of the area is below the x-axis.

Example

Find the area between the curve $y = 3x(x - 2)$, the x-axis and the lines $x = 0$ and $x = 4$

First draw a sketch: The sketch shows that the area between $x = 0$ and $x = 2$ is below the x-axis and the area between $x = 2$ and $x = 4$ is above the x-axis. Therefore we need to find each area separately.

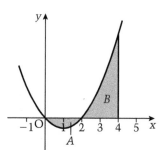

$$A = \int_0^2 (3x^2 - 6x) \, dx = [x^3 - 3x^2]_0^2 = (8 - 12) - 0 = -4$$

$$B = \int_2^4 (3x^2 - 6x) \, dx = [x^3 - 3x^2]_2^4 = (64 - 48) - (8 - 12) = 20$$

Therefore the area required $= 20 + 4 = 24$ square units.

Note that $\int_0^4 (3x^2 - 6x) \, dx$ gives the value of the area of B minus the area of A.

Exercise 3.22a

1 **(a)** Sketch the curve $y = x(x - 1)(x - 2)$.
   (b) Find the area enclosed between this curve and the x-axis.

2 **(a)** Sketch the curve $y = x^2 - 1$.
   (b) Find the area enclosed by this curve and the x-axis.

Area between two curves

When you need to find the area between two curves, you should draw a diagram. This will show any areas that may be below the x-axis and give an idea where the points of intersection are. There are two methods, which are shown in the next example.

Example

Find the area between the curves $y = x^2$ and $y = 3x(2 - x)$

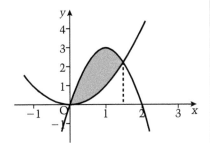

The curves intersect where $x^2 = 3x(2 - x)$

$\Rightarrow \qquad 4x^2 - 6x = 0$

$\Rightarrow \qquad x(2x - 3) = 0 \Rightarrow x = 0$ or $x = 1\frac{1}{2}$

First method

The area between $y = 3x(2 - x)$, the x-axis and $x = 1\frac{1}{2}$

is given by $\int_0^{1\frac{1}{2}} (6x - 3x^2)\, dx = \left[3x^2 - x^3\right]_0^{1\frac{1}{2}} = \frac{27}{4} - \frac{27}{8} - 0 = \frac{27}{8}$

The area between $y = x^2$, the x-axis and $x = 1\frac{1}{2}$ is given by

$\int_0^{1\frac{1}{2}} x^2\, dx = \left[\frac{1}{3}x^3\right]_0^{1\frac{1}{2}} = \frac{9}{8} - 0 = \frac{9}{8}$

Therefore the area between the curves is $\frac{27}{8} - \frac{9}{8} = \frac{9}{4}$ square units

Second method

A vertical strip in the area has length $(y = 6x - 3x^2) - (y = x^2) = 6x - 4x^2$

Therefore the area between the curves is $\int_0^{1\frac{1}{2}} (6x - 4x^2)\, dx = \left[3x^2 - \frac{4}{3}x^3\right]_0^{1\frac{1}{2}} = \frac{27}{4} - \frac{9}{2} - 0 = \frac{9}{4}$ square units

Exercise 3.22b

1 Find the area between the curves $y = 2 - x^2$ and $y = 2 - x$

2 Find the area enclosed by the y-axis and the curves $y = x^3$ and $y = 16 - x^3$

3 Find the area enclosed by the curve $y = (x - 2)(x + 2)$ and the line $y = 3x$

3.23 Volumes of revolution

Learning outcomes

- To calculate the volume formed when part of a curve is rotated round the *x*-axis

You need to know

- How to integrate simple functions
- How to evaluate a definite integral
- The formula for the volume of a cylinder

Volume of revolution

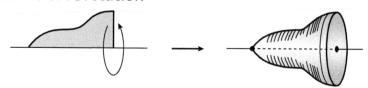

When an area is rotated about a straight line, the three-dimensional object formed is called a *solid of revolution*.

The volume formed is called a *volume of revolution*.

The line about which rotation takes place is an axis of symmetry of the solid of revolution.

All cross-sections of the solid that are perpendicular to the axis of rotation are circular.

The diagram shows the solid of revolution formed when the shaded area is rotated completely about the *x*-axis.

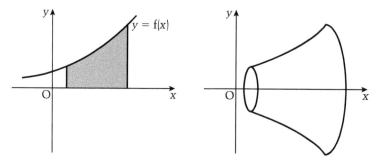

To calculate the volume of this solid we divide it into 'slices' perpendicular to the axis of rotation.

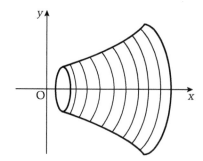

When the cuts are close together, each slice is approximately a cylinder.

When one cut is through the point P(x, y) and the volume of the solid up to this cut is V, and then another cut is made at a distance δx from the first, the volume of this slice is approximately a cylinder.

The volume of this slice is a small increase, δV, of the volume V.

This 'cylinder' has radius y and depth δx.

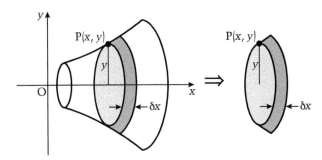

Therefore $\delta V \approx \pi y^2 \delta x$

$\Rightarrow \qquad \dfrac{\delta V}{\delta x} \approx \pi y^2$

This approximation improves as $\delta x \to 0$,

i.e. $\quad \lim\limits_{\delta x \to 0} \dfrac{\delta V}{\delta x} = \pi y^2 \quad \Rightarrow \quad \dfrac{dV}{dx} = \pi y^2$

Therefore $\quad V = \int \pi y^2 \, dx$

When the area between a curve $y = f(x)$, the x-axis, $x = a$ and $x = b$ is rotated completely about the x-axis the volume of the solid formed is given by the definite integral $V = \displaystyle\int_a^b \pi y^2 \, dx$

Example

Find the volume generated when the area between the curve $y = \sqrt{3x}$, the x-axis and the line $x = 2$ is rotated completely about the x-axis.

When we integrate with respect to x, the limits of the integration must be values of x.

$$V = \int_0^2 \pi y^2 \, dx = \pi \int_0^2 (\sqrt{3x})^2 \, dx$$

$$= \pi \int_0^2 3x \, dx = \pi \left[\tfrac{3}{2} x^2 \right]_0^2 = \pi(6 - 0) = 6\pi$$

Therefore the volume required is 6π cubic units.

Note that values for volumes of revolution are usually given in terms of π.

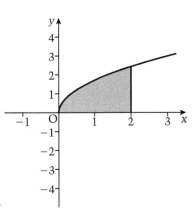

Exercise 3.23

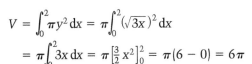

1 The area enclosed by the curve $y = 4 - x^2$ and the x-axis is rotated completely about the x-axis. Find the volume of the solid generated.

2 The area enclosed by the curve $y = x^2$, the x-axis and the lines $x = -1$ and $x = 1$ is rotated completely about the x-axis. Find the volume of the solid generated.

Learning outcomes

- To find the volume generated when a section of a curve is rotated about the *y*-axis

- To find the volume generated when the area between two curves is rotated about the *x*-axis or the *y*-axis

You need to know

- How to integrate simple functions

- How to sketch curves

- How to find the points of intersection of two curves

- The formula for the volume of a cone

Rotation about the *y*-axis

When an area is rotated about the *y*-axis, the volume formed is calculated in a similar way to rotation about the *x*-axis.

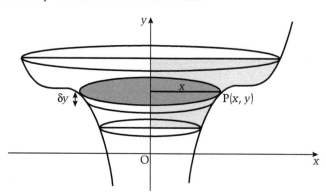

Two slices parallel to the *x*-axis, a distance δy apart, form an approximate cylinder of radius *x* and depth δy.

The volume of this cylinder is $\pi x^2 \, \delta y$.

$$\therefore \quad \frac{\delta V}{\delta y} \approx \pi x^2 \quad \Rightarrow \quad \frac{dV}{dy} = \pi x^2 \quad \Rightarrow \quad V = \int \pi x^2 \, \delta y$$

Therefore the volume generated when the part of the curve $y = f(x)$ between $y = a$ and $y = b$ is rotated completely about the *y*-axis is

$$\text{given by } V = \int_a^b \pi x^2 \, dy$$

Note that $\int \dots dy$ means that the integration has to be done with respect to *y*, so the function to be integrated must be in terms of *y* and the limits must be values of *y*.

Example

The region defined by the inequalities $y \geqslant x^2 + 2$, $2 \leqslant y \leqslant 3$ is rotated about the *y*-axis.

Find the volume of the solid generated.

The equation of the curve is $y = x^2 + 2$

$V = \int_2^3 \pi x^2 \, dy$ with $y = x^2 + 2$,

i.e. $x^2 = y - 2$, gives

$$V = \pi \int_2^3 (y - 2) \, dy = \pi \left[\tfrac{1}{2} y^2 - 2y \right]_2^3$$

$$= \pi \left\{ \left(\tfrac{9}{2} - 6 \right) - (2 - 4) \right\} = \tfrac{1}{2} \pi$$

Therefore the volume generated is $\tfrac{1}{2} \pi$ cubic units.

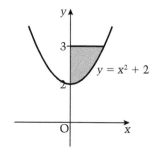

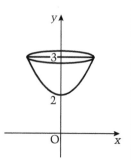

Exercise 3.24a

1 Find the volume generated when the area enclosed by $y = 9 - x^2$ and the x-axis is rotated completely about the y-axis.

2 The region enclosed by $y = x^3$, the y-axis and the line $y = 1$ is rotated completely about the y-axis. Find the volume generated.

Rotation of an area between two curves

When an area between two curves is rotated about an axis, the solid formed has a hollow section. We can find the volume of this solid by subtracting the volume formed by rotation of the inner curve from the volume formed by rotation of the outer curve,

i.e. if $y_1 = f(x)$ is the equation of the outer curve and $y_2 = g(x)$ is the equation of the inner curve, then the volume between them is given by

$$\pi \int_a^b y_1^2 \, dx - \pi \int_a^b y_2^2 \, dx \text{ where } a \text{ and } b \text{ are the values of } x \text{ where the}$$

curves intersect.

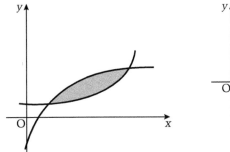 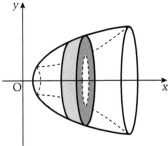

An alternative method may simplify the working. This involves just one integral.

A slice through the solid gives a shape whose cross-section is an ***annulus***. (An annulus is the area between two concentric circles.)

If, for a value of x, y_1 is the y-coordinate of a point on one of the curves and y_2 the corresponding point on the other curve, where $y_1 > y_2$, then the area of the cross-section is $\pi(y_1^2 - y_2^2)$

The volume of a slice of thickness δx is then $\pi(y_1^2 - y_2^2) \, \delta x$

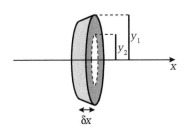

Therefore $\delta V = \pi(y_1^2 - y_2^2) \, \delta x$

$$\Rightarrow \quad \frac{\delta V}{\delta x} \approx \pi(y_1^2 - y_2^2) \quad \Rightarrow \quad \frac{dV}{dx} = \pi(y_1^2 - y_2^2)$$

$$\therefore \quad V = \pi \int (y_1^2 - y_2^2) \, \delta x$$

This is useful when the equations of the curves are similar. For example,

when $y_1 = \frac{1}{x} + 1$ and $y_2 = \frac{1}{x}$, then $y_1^2 - y_2^2 = \left(\frac{1}{x} + 1\right)^2 - \left(\frac{1}{x}\right)^2 = \frac{2}{x} + 1$

Each problem should be assessed to determine the best method.
A sketch of the curves involved will help you do this.

Example

Find the volume of the solid formed when the area enclosed by the curve $y = \sqrt{x-1}$ and the line $2y = x - 1$ is rotated completely about the x-axis.

First find where the curve and line intersect.

$$\sqrt{x-1} = \tfrac{1}{2}(x-1)$$

$\Rightarrow \quad x - 1 = \tfrac{1}{4}(x-1)^2$

$\Rightarrow \quad (x-1)^2 - 4(x-1) = 0$

$\Rightarrow \quad (x-1)(x-1-4) = 0$

$\Rightarrow \quad (x-1)(x-5) = 0$

$\Rightarrow \quad x = 1 \ \text{ or } \ x = 5$

When $x = 1$, $y = 0$ and when $x = 5$, $y = 2$

Therefore the graphs intersect at $(1, 0)$ and $(5, 2)$

Next sketch the graphs.

The hollow section formed by rotating the line completely about the x-axis is a cone, and the volume of the cone can be found without integration.

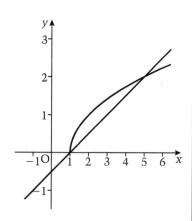

This cone has base radius 2 units and height 4 units.

Therefore the volume of the hollow cone is $\frac{16}{3}\pi$

$\left(\text{using the formula } V = \tfrac{1}{3}\pi r^2 h\right).$

The volume generated when the section of the curve $y = \sqrt{x-1}$ between $x = 1$ and $x = 5$ is rotated completely about the x-axis is given by

$$V = \pi \int_{1}^{5} (x-1)\,dx = \pi \left[\tfrac{1}{2}x^2 - x\right]_{1}^{5}$$

$$= \pi\left\{\left(\tfrac{25}{2} - 5\right) - \left(\tfrac{1}{2} - 1\right)\right\} = 8\pi$$

Now $8\pi - \frac{16}{3}\pi = \frac{8}{3}\pi$

Therefore the required volume is $\frac{8}{3}\pi$ cubic units.

Example

The diagram shows the area enclosed by the line $y = x + 1$, the curve $y = \sqrt{x - 1}$, the x- and y-axes and the line $x = 4$

Find the volume generated when this area is rotated completely about the x-axis.

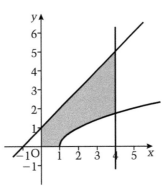

This volume needs to be calculated in two separate calculations as the limits are different for the volume generated by the rotation of the line and the volume generated by the rotation of the curve.

The volume generated when the section of the line $y = x + 1$ between $x = 0$ and $x = 4$ is rotated about the x-axis is

$$\pi \int_0^4 (x + 1)^2 \, dx = \pi \int_0^4 (x^2 + 2x + 1) \, dx$$

$$= \pi \left[\frac{1}{3} x^3 + x^2 + x \right]_0^4$$

$$= \pi \left(\frac{1}{3} (4)^3 + (4)^2 + 4 - 0 \right) = \frac{124\pi}{3}$$

The volume generated when the section of the curve $y = \sqrt{x - 1}$ between $x = 1$ and $x = 4$ is rotated about the x-axis is

$$\pi \int_1^4 (\sqrt{x - 1})^2 \, dx = \pi \int_1^4 (x - 1) \, dx = \pi \left[\frac{1}{2} x^2 - x \right]_1^4$$

$$= \pi \left[\left(\frac{1}{2} (4)^2 - 4 \right) - \left(\frac{1}{2} (1)^2 - 1 \right) \right] = \frac{9\pi}{2}$$

Therefore the volume of the solid formed is

$\dfrac{124\pi}{3} - \dfrac{9\pi}{2} = \dfrac{221\pi}{6}$ cubic units.

Exercise 3.24b

1. Find the volume generated when the area between the curve $y = x^2$ and the line $y = 3x$ is rotated completely about the x-axis.

2. The area between the curves $y = x^2$, $y = x^2 + 1$ and the lines $x = 0$ and $x = 2$ is rotated about the x-axis. Find the volume generated.

3. (a) Draw a sketch showing the line $y = x - 1$ and the curve $y = (x - 1)^3$ between $x = -1$ and $x = 2$

   (b) The area enclosed by the line and the curve, between $x = 0$ and $x = 1$ is rotated completely about the x-axis. Find the volume generated.

Differential equations

Any equation with terms involving $\frac{dy}{dx}$, $\frac{d^2y}{dx^2}$, and so on, is called a **differential equation**.

An equation involving only terms in $\frac{dy}{dx}$ is called a **first order** differential equation.

For example $x + y\frac{dy}{dx} = 2$ is a first order differential equation.

If an equation involves $\frac{d^2y}{dx^2}$, it is called a **second order** differential equation.

Rates of increase

We know that $\frac{dy}{dx}$ represents the rate at which y increases with respect to x.

When the varying value of a quantity P depends on the change in another quantity Q, then the rate of increase of P with respect to Q is $\frac{dP}{dQ}$.

Such changes occur frequently in everyday life, for example, liquid expands when it is heated. When the volume of a quantity of liquid is V and the temperature is T, then the rate at which the volume of the liquid increases with respect to changing temperature can be modelled by $\frac{dV}{dT}$.

Another example is the profit, P, made by a bookseller. This depends on the number, n, of books sold (among other factors). So the rate at which profit increases as n changes can be modelled as $\frac{dP}{dn}$.

Formation of differential equations

The motion of a particle is often modelled by a differential equation. If you are studying mechanics you will be familiar with displacement, velocity and acceleration, where velocity is the rate of change of displacement with respect to time and acceleration is the rate of change of velocity with respect to time.

If an object falls through a medium that causes its velocity v to decrease with respect to time at a rate that is proportional to its velocity, then $\frac{dv}{dt}$ measures the rate of increase with respect to time, so $\frac{dv}{dt}$ is negative. As $\frac{dv}{dt}$ is proportional to v, we can model this movement with the differential equation $\frac{dv}{dt} = -kv$ where k is a constant of proportionality.

As v is the rate of change of the displacement, s, $\frac{ds}{dt} = v$, we can also model this movement with the equation $\frac{d^2s}{dt^2} = -kv$

Note that when we are told the rate at which a quantity is changing, we assume that the change is with respect to time unless we are told otherwise.

Example

Form a differential equation to model the following information.

Water is leaking from a cylindrical tank such that the rate at which the depth of water is decreasing is proportional to the volume of water left in the tank.

The rate at which the depth, h, is changing is $\frac{dh}{dt}$.

The volume of water in the tank is $\pi r^2 h$.

$\frac{dh}{dt}$ is negative as h is decreasing,

$\therefore \quad \frac{dh}{dt} \propto -\pi r^2 h$

π and r are both constants so we can write this equation as

$\frac{dh}{dt} = -kh$

Exercise 3.25

Form a differential equation to model the following data.

1 When bacteria are grown in a culture, the rate of increase of the number of cells C is proportional to the number of cells present at that time.

2 A body moving in a straight line moves so that the rate of change of its displacement, s, from a fixed point is inversely proportional to s.

3 The rate at which a cereal crop grows is such that its height, h cm, increases at a rate with respect to time which is proportional to the difference between its final height, H, and its present height.

4 Grain is being drained from a hopper. The rate of change of the volume, V, of grain in the hopper is inversely proportional to the volume of grain remaining in the hopper.

3.26 Solving differential equations

- To solve differential equations

- How to integrate simple functions
- How to find the constant of integration from given information
- How to use substitution to integrate

Solving differential equations

Solving a differential equation means finding a direct relationship between the variables.

For example, the general solution of the differential equation $\frac{dy}{dx} = 2x$ is $y = x^2 + c$

To get a unique solution, we need a pair of corresponding values of x and y.

Differential equations often involve a constant of proportionality, so the solution will involve two unknown constants. In this case we need two additional pieces of information to get a unique solution.

For example, given the differential equation $\frac{ds}{dt} = kt$, and that $s = 1$ when $t = 0$ and that $s = 6$ when $t = 10$, then

$$\frac{ds}{dt} = kt \quad \Rightarrow \quad s = \tfrac{1}{2}kt^2 + c$$

$s = 1$ when $t = 0$ gives $1 = c$

$s = 6$ when $t = 10$ gives $6 = 50k + 1 \quad \Rightarrow \quad k = \tfrac{1}{10}$

$\therefore \quad s = \tfrac{1}{10}t^2 + 1$

We also need two additional pieces of information when there is a second derivative involved in the differential equation. In this case two integrations are needed, each of which will introduce a constant.

For example, if $\frac{d^2y}{dx^2} = 3x^2$ then integrating once gives

$$\frac{dy}{dx} = x^3 + c$$

Integrating again gives
$$y = \tfrac{1}{4}x^4 + cx + k$$

If we know that $y = 5$ when $x = 2$,
this gives $\quad 5 = 4 + 2c + k$ [1]

If we also know that $y = 1$ when $x = 1$,
we have $\quad 1 = \tfrac{1}{4} + c + k$ [2]

Solving [1] and [2] simultaneously gives $c = \tfrac{1}{4}$ and $k = \tfrac{1}{2}$

Therefore the solution of the equation $\frac{d^2y}{dx^2} = 3x^2$ is

$y = \tfrac{1}{4}x^4 + \tfrac{1}{4}x + \tfrac{1}{2}$
$\Rightarrow \quad 4y = x^4 + x + 2$

To solve a differential equation, you need to know how to integrate it. You do not need to understand every detail about the situation that gives rise to the differential equation.

Example

A body moves so that at time t seconds its displacement, s metres from a fixed point O is modelled by $\frac{d^2s}{dt^2} = \sqrt{t}$. When $t = 0$, $\frac{ds}{dt} = 5$ and $s = 2$. Find the direct relationship between s and t. Hence find the value of s predicted by this model when $t = 4$

$$\frac{d^2s}{dt^2} = t^{\frac{1}{2}} \implies \frac{ds}{dt} = \frac{2}{3}t^{\frac{3}{2}} + c$$

When $t = 0$, $\frac{ds}{dt} = 5$, $\quad \therefore \quad c = 5$

$$\frac{ds}{dt} = \frac{2}{3}t^{\frac{3}{2}} + 5 \implies s = \frac{4}{15}t^{\frac{5}{2}} + 5t + k$$

When $t = 0$, $s = 2$, $\quad \therefore \quad k = 2$

Hence $\quad s = \frac{4}{15}t^{\frac{5}{2}} + 5t + 2$

When $t = 4$, $s = \frac{128}{15} + 22 = 30\frac{8}{15}$

The model predicts that $s = 30\frac{8}{15}$ when $t = 4$

Exercise 3.26a

1 The rate of change of a quantity r with respect to θ is given by $\frac{dr}{d\theta} = 3\sin\theta$. When $\theta = \frac{\pi}{3}$, $r = 2$. Find r in terms of θ.

2 The variation of a quantity P with respect to r is modelled by the differential equation $\frac{d^2P}{dr^2} = 12r^2 - 6r$. It is known that when $r = 1$, $P = 6$ and $\frac{dP}{dr} = -1$. What does this model predict that the value of P will be when $r = 3$?

Integration by separating the variables

Many differential equations used to model situations are of the form $\frac{dy}{dx} = f(y)$. We cannot integrate $f(y)$ with respect to x, so we need to change the form of the differential equation.

We know from Topic 3.18 that $\int f(u)\frac{du}{dx}\,dx = \int f(u)\,du$, where u is a function of x.

Therefore $\int f(y)\frac{dy}{dx}\,dx = \int f(y)\,dy$

This means that integrating $\left((\text{a function of } y)\frac{dy}{dx}\right)$ with respect to x, is *equivalent* to integrating (the same function of y) with respect to y.

Now we can write $\frac{dy}{dx} = f(y)$ as $\frac{1}{f(y)}\frac{dy}{dx} = 1$

then $\int\left(\frac{1}{f(y)}\frac{dy}{dx}\right)dx = \int 1\,dx$ becomes $\int\left(\frac{1}{f(y)}\right)dy = \int 1\,dx$

This is called *integration by separating the variables* because what we have effectively done in going from $\frac{dy}{dx} = f(y)$ to $\int\left(\frac{1}{f(y)}\right)dy = \int 1\,dx$ is to gather all the terms containing y on one side and all the terms containing x on the other side, i.e. we have 'separated' the numerator and denominator of $\frac{dy}{dx}$.

For example, given $\frac{dr}{dt} = \frac{2}{r}$, then multiplying by r gives $r\frac{dr}{dt} = 2$

so $\int r\,dr = \int 2\,dt \implies \frac{1}{2}r^2 = 2t + c$

Example

The atoms in a radioactive material are disintegrating at a rate that is modelled as inversely proportional to the number of atoms present at any given time, t, measured in days. Initially there are N atoms present.

(a) Form and solve a differential equation to represent this information.

(b) Half the mass disintegrates in 200 days. Find how long the model predicts that it will take for three-quarters of the mass to disintegrate.

(a) If n is the number of atoms present at any given time, then the rate of change of n is negative,

$\therefore \quad \frac{dn}{dt} = -\frac{k}{n} \implies n\frac{dn}{dt} = -k \qquad$ where k is a constant.

Therefore $\int n\,dn = -\int k\,dt$

$\implies \qquad\qquad \frac{1}{2}n^2 = -kt + c$

Initially, i.e. when $t = 0$, $n = N$, $\therefore \frac{1}{2}N^2 = c$

hence $\qquad\qquad \frac{1}{2}n^2 = -kt + \frac{1}{2}N^2$

(b) When $t = 200$, $n = \frac{1}{2}N$

$\therefore \qquad\qquad \frac{1}{2}\left(\frac{1}{2}N\right)^2 = -200k + \frac{1}{2}N^2$

$\implies 200k = \frac{3}{8}N^2 \quad$ so $\quad k = \frac{3}{1600}N^2$

i.e. $\qquad\qquad \frac{1}{2}n^2 = -\frac{3}{1600}N^2t + \frac{1}{2}N^2$

When $\frac{3}{4}$ of the mass has disintegrated, $n = \frac{1}{4}N$

$\implies \frac{1}{2}\left(\frac{1}{4}N\right)^2 = -\frac{3}{1600}N^2t + \frac{1}{2}N^2 \implies \frac{1}{32} = -\frac{3}{1600}t + \frac{1}{2}$

$\implies \quad t = 250$

The model predicts that it will take 250 days for $\frac{3}{4}$ of the mass to disintegrate.

Example

The rate of increase in the number, n, of people infected by a virus is modelled as being proportional to the square root of the number of people already infected. Nine people were infected 5 days after the first person was infected.

(a) Form and solve a differential equation to represent this information.

(b) How many days, to the nearest day, does the model predict that it will take for 100 people to be infected?

(a) When n people are infected, $\dfrac{dn}{dt} = kn^{\frac{1}{2}}$ where k is a constant.

$\therefore \int n^{-\frac{1}{2}} dn = \int k \, dt \quad \Rightarrow \quad 2n^{\frac{1}{2}} = kt + c$

When $t = 0$, $n = 1$

(This is not given explicitly but the days are counted from when the first person is infected.)

$\therefore \quad c = 2$

When $t = 5$, $n = 9 \quad \Rightarrow \quad 6 = 5k + 2 \quad \Rightarrow \quad k = \frac{4}{5}$

$\therefore \quad 2n^{\frac{1}{2}} = \frac{4}{5}t + 2$

(b) When $n = 100$, $\qquad 20 = \frac{4}{5}t + 2$

$\Rightarrow \qquad\qquad\qquad\qquad t = \frac{45}{2}$

The model predicts it will take approximately 23 days for 100 people to be infected.

Exercise 3.26b

1 The velocity, $v \, \text{m s}^{-1}$, of a ball rolling along the ground is such that, t seconds after it started, $\int v^{-\frac{1}{2}} dv = \int k \, dt$. Given that $v = 5$ when $t = 0$ and that $v = 2$ when $t = 3$, solve the differential equation to give a direct relationship between v and t.

2 Water is dripping from a tap on to a concrete surface where it is forming a circular damp patch. Two hours after the tap started dripping, the radius of the damp patch was 20 cm.

   The rate at which the radius, r cm, of the damp patch is increasing is modelled as being proportional to $\dfrac{1}{r}$.

   (a) Form and solve a differential equation giving r in terms of t, the number of hours elapsed after the tap starts to drip.

   (b) How long, to the nearest hour, does the model predict it will take for the radius of the damp patch to reach 1 m?

3 Grain is pouring from a hopper on to a barn floor where it forms a conical pile whose height h is increasing at a rate that is inversely proportional to h^3. The initial height of the pile is 2 m and the height doubles after a time T. Find, in terms of T, the time after which its height has grown to 6 m.

Section 3 Practice questions

1 Find:
   (a) $\lim\limits_{h \to 0} \dfrac{h}{\sqrt{+2} - \sqrt{2}}$
   (b) $\lim\limits_{x \to -2} \dfrac{x + 2}{x^2 - 2x - 8}$

2 (a) The function f is given by
   $$f(x) = \begin{cases} x^2 + 1, & x \geq 2, \\ 2x, & x < 2, \end{cases} \quad x \in \mathbb{R}$$
   Find $\lim\limits_{x \to 2^+} f(x)$ and $\lim\limits_{x \to 2^-} f(x)$.
   Hence explain whether or not f(x) is continuous at $x = 2$
   (b) Repeat (a) for
   $$f(x) = \begin{cases} 3x + 1, & x > 3 \\ x^2 + 1, & x \leq 3 \end{cases} \quad x \in \mathbb{R}$$

3 Differentiate from first principles:
   (a) $\sin 2x$
   (b) $\sqrt{x}$

4 Find $\dfrac{dy}{dx}$ when:
   (a) $y = (2x - 1)(3x + 2)$
   (b) $y = 4\sin x - 3\cos x$
   (c) $y = \dfrac{x^2 - 2x + 1}{x - 1}$

5 Given that $y = x\sin x$
   find $\dfrac{dy}{dx}$ and $\dfrac{d^2y}{dx^2}$.

6 Find f'(x) when f(x) is:
   (a) $\dfrac{x}{x^2 - 1}$
   (b) $x\sqrt{x + 1}$
   (c) $\dfrac{x}{\sin x}$
   (d) $\sqrt{x^2 + 1}$
   (e) $\cos\left(2x - \dfrac{\pi}{3}\right) \sin\left(2x - \dfrac{\pi}{3}\right)$

7 Find $\dfrac{dy}{dx}$ in terms of t when:
   (a) $y = 2t, \ x = t^2 - 2t$
   (b) $y = 3\cos t, \ x = 4 - 5\sin t$

8 The equation of a curve is
   $$y = 3(x - 5)^3$$
   A point is moving along the curve so that x is increasing at the constant rate of 0.2 cm per second. Find the rate of change of y when x is 1.5 cm.

9 A spherical balloon is losing air at the rate of 0.5 cm^3 per second.
   Find the rate of change of the radius of the balloon when the radius is 20 cm.
   $\left(\text{The volume of a sphere of radius } r \text{ is } \frac{4}{3}\pi r^3.\right)$

10 Find the range of values of x for which the function given by $f(x) = \dfrac{4x}{(2 - x)^2}$ is increasing.

11 Show that
   $$y = \dfrac{x^2}{(3x - 3)^2}$$
   has one stationary value and find it.

12 Find the stationary points on the curve
   $$y = 3x^4 - 4x^3 - 6x^2 + 12x - 5$$
   and distinguish between them.

13 The curve $y = ax^3 - x^2 + b$ has a maximum value of 4 when $x = 0$ and a minimum value c when $x = 2$.
   Find the values of a, b and c.

14 Sketch the curve whose equation is
   $$y = \dfrac{4x}{(2 - x)^2}$$
   (You can use your results from question 10.)

15 Find the equation of the curve which goes through the point $\left(\dfrac{\pi}{2}, 1\right)$ and for which
   $y = \int (5\cos \theta)\, d\theta$

16 A curve passes through the points (0, 1) and (1, 1). The equation of the curve is such that
   $\dfrac{d^2y}{dx^2} = 4 - 6x$
   Find the equation of the curve.

17 (a) Use the substitution $u = \sin 2x$ to find
   $\int \cos 2x \sin^2 2x\, dx$
   (b) Use the substitution $u = x^2 - 1$ to find
   $\int 6x(x^2 - 1)^4\, dx$

18 Evaluate:
   (a) $\displaystyle\int_2^4 (3x - 4)\, dx$ (b) $\displaystyle\int_{\frac{\pi}{2}}^{\pi} 2\cos \theta\, d\theta$

19 (a) Use the substitution $u = x^2 - 1$ to evaluate
   $\displaystyle\int_1^2 \left(x\sqrt{x^2 - 1}\right) dx$
   (b) Use the substitution $u = \sin \theta$ to evaluate
   $\displaystyle\int_0^{\frac{\pi}{4}} \cos \theta\, (\sin \theta - 1)^3\, d\theta$

20 (a) Find the area enclosed by the curve $y = 4 - x^2$ and the x-axis.

(b) Find the area enclosed by the curve $y = x^3$, the y-axis and the lines $y = 1$ and $y = 2$

21 (a) Sketch the curve
$$y = (x + 1)(x - 1)(x + 2)$$

(b) Find the area enclosed by this curve and the x-axis.

22 Find the area enclosed by the curves $y = x^2 + 1$ and $y = 5 - x^2$

23 Find the volume generated when the area enclosed by the curve $y = x^3 + 1$, the y-axis, the x-axis and the line $x = 1$ is rotated completely about the x-axis.
Give your answer in terms of π.

24 (a) Find the equation of the tangent to the curve $y = x^2 + 2$ at the point on the curve where $x = 2$

(b) Draw a sketch to show the area enclosed between the curve and tangent in part **(a)** and the x- and y-axes.

(c) Find the volume generated when the area described in part **(b)** is rotated completely about:

(i) the x-axis

(ii) the y-axis.

25 Solve the differential equation
$$\frac{dy}{dx} = 6y^2$$
given that $y = 3$ when $x = 1$

26 Solve the differential equation
$$\frac{dy}{dx} = \frac{x}{y}$$
given that $y = -3$ when $x = 2$

27 Air is escaping from a spherical balloon at a rate that is proportional to V^2, where $V\,cm^3$ is the volume of the balloon.

(a) Use the information above to form a differential equation in terms of V and t where t seconds is the time in seconds that has elapsed from when the air started to escape.

(b) The initial radius of the balloon was $10\,cm$. Ten seconds after air started to escape, the radius, $r\,cm$, of the balloon was $5\,cm$.
How long will it be before the radius of the balloon is $2\,cm$?
(The volume of a sphere is $\frac{4}{3}\pi r^3$.)

28 Given that
$$y = 3\cos 2x,$$
find $\frac{d^2y}{dx^2}$ in terms of y.

29 The point $\left(\frac{2}{3}, \frac{44}{9}\right)$ is a point of inflexion on the curve
$$y = ax^3 - 6x^2 + bx + 4$$
Find the equation of the tangent to the curve at the point where $x = 1$

30 $\displaystyle\int_a^1 \frac{1}{x^2}\,dx = 10$ where $a > 0$. Find the value of a.

31 Solve the differential equation
$$\frac{d^2y}{dx^2} = 6x + 4$$
given that, when $x = 0$, $\frac{dy}{dx} = 3$ and $y = 9$

32 Find the volume generated when the area between the curves
$$y = x^2 \quad\text{and}\quad y = 8 - x^2$$
is rotated completely about the x-axis.

33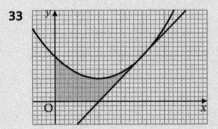

The diagram shows the area enclosed by the curve $y = x^2 - 2x + 2$, the x- and y-axes and the tangent to the curve at the point where $x = 2$

(a) Find this area.

(b) Find the volume generated when this area is rotated completely about the x-axis.

Index